Molecular Biology Intelligence Unit

HOX Gene Expression

Spyros Papageorgiou, Ph.D.
Institute of Biology
NRC 'Demokritos'
Athens, Greece

Landes Bioscience / Eurekah.com
Austin, Texas
U.S.A.

Springer Science+Business Media
New York, New York
U.S.A.

HOX Gene Expression

Molecular Biology Intelligence Unit

Landes Bioscience / Eurekah.com
Springer Science+Business Media, LLC

ISBN: 978-0-387-68989-0 Printed on acid-free paper.

The choice of italics or case for Hox is determined by the authors, preference and may differ between chapters.

Springer Science+Business Media, LLC, 233 Spring Street, New York, New York 10013, U.S.A.
http://www.springer.com

Please address all inquiries to the Publishers:
Landes Bioscience / Eurekah.com, 810 South Church Street, Georgetown, Texas 78626, U.S.A.
Phone: 512/ 863 7762; FAX: 512/ 863 0081
http://www.eurekah.com
http://www.landesbioscience.com

Printed in the United States of America.

9 8 7 6 5 4 3 2 1

Library of Congress Cataloging-in-Publication Data

A C.I.P. Catalogue record for this book is available from the Library of Congress.

CONTENTS

The choice of italics or case for Hox is determined by the authors, preference and may differ between chapters.

EDITOR

Spyros Papageorgiou, Ph.D.
Institute of Biology
NRC 'Demokritos'
Athens, Greece
Email: spapage@bio.demokritos.gr
Appendix

CONTRIBUTORS

Susan V. Bryant
Department of Developmental
and Cell Biology
The Developmental Biology Center
University of California Irvine
Irvine, California, USA
Chapter 7

Clemente Cillo
Department of Clinical
and Experimental Medicine
Federico II University Medical School
Naples, Italy
Email: clecillo@unina.it
Chapter 9

Mark Featherstone
School of Biological Sciences
Nanyang Technological University
Singapore, Republic of Singapore
Email: mark.featherstone@mcgill.ca
Chapter 6

David E. K. Ferrier
Department of Zoology
University of Oxford
Oxford, U.K.
Email: david.ferrier@zoo.ox.ac.uk
Chapter 4

David M. Gardiner
Department of Developmental
and Cell Biology
The Developmental Biology Center
University of California Irvine
Irvine, California, USA
Email: dmgardin@uci.edu
Chapter 7

Anthony Gavalas
Foundation for Biomedical Research
of the Academy of Athens
Athens, Hellas/Greece
Email: agavalas@bioacademy.gr
Chapter 8

Walter J. Gehring
Department of Cell Biology
University of Basel
Basel, Switzerland
Email: walter.gehrig@unibas.ch
Chapter 1

Mina Gouti
Foundation for Biomedical Research
of the Academy of Athens
Athens, Hellas/Greece
Chapter 8

Laila Kobrossy
Department of Molecular and Cellular
Biology
Harvard University
Cambridge, Massachusetts, USA
Chapter 6

Robb Krumlauf
Stowers Institute for Medical Research
Kansas City, Missouri, USA
Email: rek@stowers-institute.org
Chapter 2

Christof Nolte
Stowers Institute for Medical Research
Kansas City, Missouri, USA
Chapter 2

Sonja J. Prohaska
Department of Computer Science
and Engineering
Arizona State University
Phoenix, Arizona, USA
Chapter 5

Peter F. Stadler
Lehrstuhl für Bioinformatik
Institut für Informatik
Universität Leipzig
Leipzig, Germany
Email: studla@bioinf.uni-leipzig.de
Chapter 5

Cheryll Tickle
Division of Cell and Developmental
Biology
School of Life Sciences
University of Dundee
Dundee, Scotland, U.K.
Email: c.a.tickle@dundee.ac.uk
Chapter 3

Günter P. Wagner
Barmhertige Brüder-Hospital
A-1020 Wien, Austria
Email: gunter.wagner@yale.edu
Chapter 5

PREFACE

This book is about an old fundamental problem of embryogenesis and the contemporary methods and ideas used to approach this subject. In the long history of science the invention of a new instrument or an innovative technique have played an unpredictable role with far reaching consequences for scientific progress and society in general. Such an outstanding case is the construction of the pendulum by Galileo who used it as a 'clock' in order to measure time with unprecedented accuracy for the standards of his era. With this new instrument he could precisely determine the kinematic features of moving bodies and lay down the foundations of the classical mechanics that was later developed by Newton. During the last decade or so we have experienced another case of the unexpected consequences stemming from the exploitation of a novel invention: the need of some collaborating laboratories of CERN to exchange their collected results led to the use of telecommunication cables and the data, in the form of electronic documents, which were transmitted and shared by all via a computer network and suitable browsing software. This facility initiated the spectacular development of the WWW which revolutionized not only scientific advancement but most social activities of our times. The present book with its online edition is an example of the profound changes introduced recently in the area of communications in general and in publishing in particular.

In biology an analogous revolution occurred during the last three decades when some ingenious biochemical methods and techniques were developed, in particular gene cloning and sequencing. Developmental and evolutionary biology have particularly benefitted from these innovations and the questions that can now be addressed and answered go to an unprecedented depth. The case of *Hox* genes is a typical example of how a longstanding morphological enigma can now be tackled at its molecular and genetic roots. In 1894 Bateson systematically analyzed for the first time in fruitflies mutations that he termed *homeotic mutations,* in which one body region develops at the location of another body region. For many decades classical genetics and embryological observations were the only means employed in the study of these peculiar mutations. Nevertheless, hard work, experimental skills and ingenuity resulted in the accumulation of important observations which are indispensable for the formulation of some illuminating empirical rules. This effort was rewarded in 1978 when E.B. Lewis established a physical correlation between the extent of the functional domains of the *Bithorax* genes of the fruitfly and the location of these genes in the third chromosome. This unexpected property, coined collinearity, was later found to be not only a feature of the *Bithorax* complex in insects but a developmental strategy that controls axial patterning of most animal species.

With the discovery of the homeobox in the early 80s, the era of molecular embryonic development came dynamically into play imposing a complete metamorphosis of the field. Most efforts are now focused on clarifying the structure, regulation and interactions of genes involved in different stages of development. In this pursuit the study of *Hox* genes plays a central role.

This book is a synopsis of the activity of homeotic genes without aiming to cover all features and related topics of this wide area of research. For instance the role of homeobox genes in plant development is not included and only selected aspects of Hox gene expression in specific examples of vertebrate organogenesis are presented in detail. Since the book is not intended to be all-encompassing, the choice of chapter subjects is selective. Nevertheless I hope the choice is coherent and representative of most research directions where Hox genes are involved. It is only just the beginning of the molecular era of biology, but it is an exciting time because one feels that the discoveries will enrich our knowledge will converge from different scientific disciplines.

I would like to take this opportunity to thank the authors of this volume who accepted promptly to contribute and succeeded in giving a balanced and authoritative overview of this rapidly progressing field. As mentioned above, it is a great advantage of electronic publishing that the chapters appear online and they are stored in the Eurekah updatable database so that a continuous flow of new results may be incorporated in the book. The idea for creating this book is due to Dr. Ronald Landes, the publisher of Landes Bioscience. I hope his initiative will prove helpful to many graduate students and researchers working in different branches of biology and medicine. Thanks are also due to Ms. Cynthia Conomos, the Book Publications Director, who undertook with enthusiasm the administrative responsibilities of the edition.

Spyros Papageorgiou

CHAPTER 1

The Homeobox as a Key for Understanding the Principles of the Genetic Control of Development

Walter J. Gehring*

Abstract

The discovery of the homeobox, and the Hox gene clusters have uncovered a general universal principle of the genetic control of development. In all bilaterian animals these Hox clusters determine the body plan along the antero-posterior axis. Despite a bewildering diver-sity of modes of development, ranging from animals with a fixed cell lineage and a predetermined egg architecture to organisms with a highly variable cell lineage whose development is primarily based on cellular interactions, all metazoans share Hox gene clusters, characterized by a high degree of sequence conservation and a colinear gene arrangement in which the Hox genes are arranged on the chromosome in the same order as they are expressed along the antero-posterior axis of the developing embryo. This arrangement can only be understood on the basis of evolution. Hox genes have provided the entry point for a newly emerging field, evolutionary developmental genetics.

Introduction

The importance of genes in controlling development was first pointed out by T. H. Morgan who put forward the hypothesis that development is controlled by differential gene activity.[1] The first homeotic gene was identified in Drosophila by Bridges as early as 1915 in Morgan's laboratory.[2] Mutations at the *bithorax* (*bx*) locus partially convert the halteres into wings and alter the body plan, by transforming a structure of the third thoracic segment into the corresponding structure of the second thoracic segment. E. B. Lewis began his analysis of the Bithorax-Complex (BX-C) not with the intention to study homeosis, but rather to test the hypothesis of Bridges, that the genome contains naturally occurring gene duplications, often tandemly arrayed and visible as doublet bands in the giant polytene chromosomes. Lewis thought that the *bx* mutants might represent such a case. This hypothesis later proved to be correct. However, in his 1963 paper[3] he introduced the concept of developmental genetic control which he pursued relentlessly until his death. In 1978[4] Lewis published his classical paper in *Nature* in which he proposed a model which provided a frame work for the subsequent molecular genetic analysis of homeotic genes.

The "Lewis Model" can be summarized as follows:

1. Dipteran flies have evolved from more primitive insects which had four wings, and insects have evolved from more primitive arthropods which had legs on all of their abdominal

*Walter J. Gehring—Biozentrum, Department of Cell Biology, University of Basel, Klingelbergstrasse 70, CH- 4056 Basel, Switzerland. Email: walter.gehring@unibas.ch

HOX Gene Expression, edited by Spyros Papageorgiou. ©2007 Landes Bioscience and Springer Science+Business Media.

segments. During evolution two major groups of genes must have arisen: "leg-suppressing" genes, which removed the legs from all the abdominal segments, leaving only three pairs of legs on the three thoracic segments, and "haltere-promoting" genes which suppressed the second pair of wings found in the four-winged ancestors of the flies. Loss-of-function mutations in these genes should produce flies with additional legs on the abdomen and four-winged flies respectively. Indeed, Lewis has constructed such 8-legged and four-winged flies (Fig. 1), by combining different loss-of-function mutants of the Bithorax-Complex.

2. These genes form a single cluster of pseudoallelic genes known as the Bithorax-Complex (BX-C) consisting of a battery of closely linked homeotic genes which arose by tandem duplications in the course of evolution and subsequently diverged from each other by mutation.
3. One gene function is required for every body segment. Each gene is primarily expressed in the one segment which it specifies, and to some extent also in all segments posterior to it, so that all genes of the BX-C are expressed in the last abdominal segment. Each segment is specified by an array of homeotic genes in a combinatorial fashion.
4. The primitive or ground state is the mesothoracic segment T2. Loss-of-function mutations lead towards the ground state (e.g., the four-winged fly: T3→T2), whereas gain-of-function mutants like *Contrabithorax* or *Haltere-mimic* lead away from the ground state to flies with four halteres (T2→T3).
5. The genes of the Bx-C are arranged in the same order on the chromosome as they are expressed in the segments of the embryo. This is designated as the "colinearity rule".

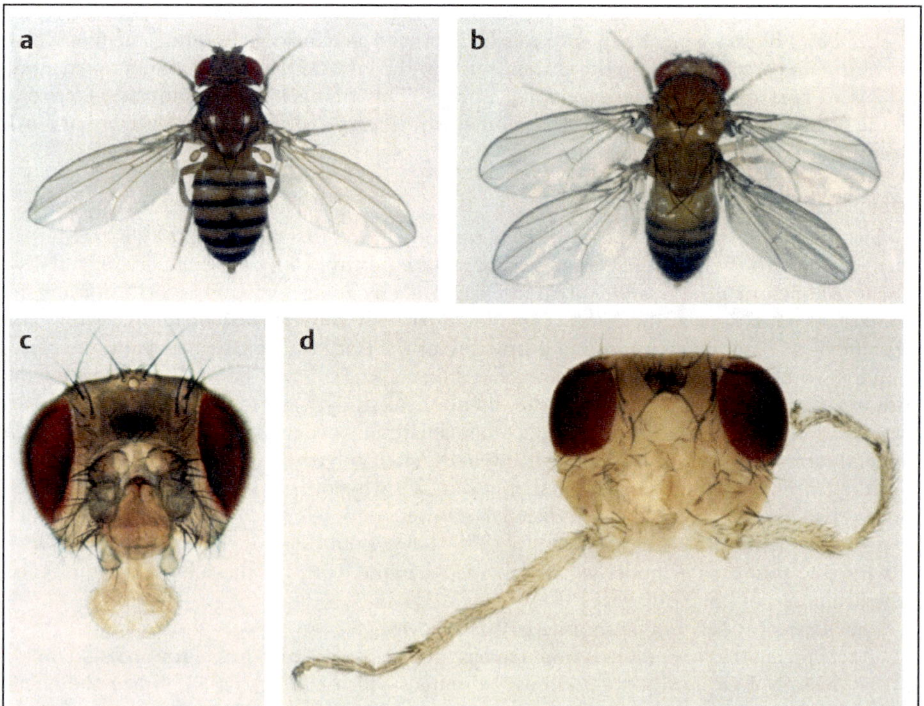

Figure 1. Homeotic mutants in Drosophila: a) Wildtype fly; b) Four-winged fly constructed by E.B. Lewis; c) Head of a wildtype fly; d) Antenna-to-leg transformation in an Antennapedia mutant ($Antp^{Ns}/Antp^{73b}$).

6. Finally Lewis assumed that "the various BX-C substances acted indirectly by repressing or activating other sets of genes which then directly determine the specific structures and functions that characterize given segment". This implies that the genes of the BX-C are regulatory genes whose function it is to regulate the activity of their target genes.

On the basis of classical genetics two major modifications of this model had to be introduced:

1. The discovery of compartments by Morata, Ripoll and Garcia-Bellido[5] indicated that the expression of the genes of the BX-C was not segmental but rather parasegmental, which explains the earlier observation that *bithorax* (*bx*) mutations transform only the anterior part of the haltere into wing structures, whereas *bithoraxoid* (*bxd*) mutations transform only the posterior part of the haltere.
2. Standard chemical mutagenesis of the BX-C by Morata and collaborators[6] showed that there are only three essential genes in the BX-C, *Ultrabithorax* (*Ubx*), *abdominal-A* (*abd-A*) and *Abdominal-B* (*Abd-B*), and not one for every segment. However, the molecular analysis later showed that the genes identified functionally by Lewis correspond to (para)segment-specific cis-regulatory regions (enhancers)[7] or to genes encoding regulatory micro-RNAs.[8]

The Lewis model invited a molecular analysis, but there was no clue as to the biochemical nature of the BX-C gene products and the cloning of homeotic genes had to await methods for cloning of genes without any information about the molecular nature of their gene products. An ingenious method for positional cloning, called "walking along the chromosome" was developed by David Hogness and his collaborators and lead to the cloning of the BX-C.[9]

Molecular Cloning of the Homeotic Bithorax-Complex in Drosophila

In an extensive chromosomal walk the Hogness' group[10] first cloned 195 kb of chromosomal DNA covering the left half of the complex and mapped the known mutations in the Ultrabithorax-region. It was possible to localize the genes by restriction enzyme mapping since most of the x-ray induced mutations are due to chromosome rearrangements and therefore, can be mapped on the cloned DNA. Also, most of the spontaneous mutations were shown to be caused by insertions of mobile transposable elements rather than point mutations, in particular by the *gypsy* element. The *Ultrabithorax* (*Ubx*) locus spans some 73 kb of chromosomal DNA and comprises all the recessive loss-of-function mutants identified by Lewis, *anterobithorax* (*abx*), *bithorax* (*bx*), *bithoraxoid* (*bxd*) and *postbithorax* (*pbx*). This was confirmed by isolating cDNA clones which indicate that the exons are spread out over the entire 73 kb *Ubx* region. Therefore, *Ubx* is a very large and highly complex gene. All of the recessive mutations are loss-of-function mutants leading towards the T2 ground state and the dominant Ubx alleles are also loss-of-function mutations, since they are recessive in triploids. However, the dominant *Contraxbithorax* (*Cbx*) mutation was classified by Lewis as gain-of-function since it transformes wings into halteres (T2→T3), i.e., away from the T2 ground state. *Cbx* arose after X-irradiation together with *postbithorax* (*pbx*) and was subsequently separated by recombination. The molecular analysis indicated that a DNA segment of 17 kb from -3 kb to +14 kb is deleted in *pbx* and inserted at position -44 kb with its orientation inverted. This complicated rearrangement (transposition plus inversion), did not allow to unequivocally explain the cause of the gain-of-function mutant phenotype. This issue was settled later in *Antennapedia* by constructing gain-of-function mutants (see below).

The right half of the Bx-C was later cloned by Karch et al[11] covering another 215 kb of chromosomal DNA, the entire Bx-C comprising more than 300 kb. As predicted by Morata and collaborators there are only three protein coding genes, *Ubx*, *abdominal-A* (*abd A*) and *Abdominal-B* (*Abd B*), all which were shown to contain homeoboxes (see below). These three genes are associated with large cis-regulatory regions. Many of the "genes" identified by Lewis through mutations turned out to be segment-specific cis-regulatory elements.[7] However, some of these intergenic regions are extensively transcribed. In particular a set of *bithoraxoid* transcripts of 1.1 to 1.3 kb in size are produced by differential splicing of a series of exons derived from a 26 kb primary transcript, but these RNAs do not possess significant protein coding potential, suggesting that they might encode regulatory RNAs. Also, the *infraabdominal-4* (*iab-4*) region is transcribed.[7] Recently, it has been found that a group of Drosophila microRNAs (miRNAs) map to a hairpin located at the distal end of the *iab-4* locus[12] between the *abd-A* and *Abd-B* genes. Two miRNAs were cloned from both arms of this hairpin and designated as iab-4 -5p and iab-4 -3p.[12] iab 4 -5p selectively inhibits *Ubx* by binding to the Ubx-3'UTR which contains predicted target sites complementary to miR- iab-4 -5p.[7] The expression of a GFP-Ubx-3'UTR "sensor" transgene is repressed by ectopic expression of a *mir-iab4* mini gene. By expressing this mir-iab-4 mini gene in the haltere imaginal discs, a bonafide homeotic transformation of halteres into wings can be induced. Thus, *iab-4* encodes a miRNA and represents a bonafide homeotic regulatory gene.[8]

Molecular Cloning of Antennapedia Complex and the Discovery of the Homeobox

As a graduate student I discovered a dominant mutation which transforms the antennae and parts of the head capsule into complete middle legs and the sternopleura, the ventral part of the thorax where the legs are inserted. I called this mutation *Nasobemia*[13] and I could map it by recombination to locus 48 on the third chromosome, since it was not associated with a chromosomal inversion. A highly similar mutation, which however did not form a sternopleura on the head, had been described by Le Calvez[14] who thought that it was a dominant allele of *aristapedia* which transforms only the distal part of the antenna into tarsal structures. However, he could not map it, since it is caused by a chromosomal inversion which suppresses recombination. Around the same time Yu in E.B. Lewis' laboratory[15] induced a similar mutation by X-irradiation. His mutation which he named *Antennapedia* (*Antp*) was also associated with chromosome rearrangements with four breakpoints, one of which roughly coincided with the one in the inversion of Le Calvez in chromosome section 84 A. Since 84 A roughly corresponds to the genetic locus 48 which is far away from *aristapedia*, this suggests that all three mutations were *Antp* rather than *aristapedia* mutations. However, all the known *Antp* mutations at that time were homozygous lethal and complemented *Nasobemia* (*Ns*) for lethality. The transheterozygotes Antp/Ns have spectacular antenna to leg transformations (Fig. 1), but they are viable. Therefore, I could not conclude whether Ns and Antp were alleles, or whether they were closely linked genes with a similar phenotype. This question was resolved much later by molecular genetics,[16] and it became clear that Ns and Antp are allelic.

Following in the food steps of David Hogness my group embarked upon a chromosome walk to clone my favourite *Antennapedia* (*Antp*) gene. When Richard Garber, Atsushi Kuroiwa and I started the chromosome walk, we only knew that the *Antp* gene maps cytologically at 84 A (later assigned by Kaufman to 84 B). We started with a genomic DNA fragment from 84 F and used an inversion to "jump" into 84 B, close to the *Antp* gene. From this point we continued to walk towards the centromere for over 200 kb and mapped the breakpoints of several *Antp* inversions and deletions. Subsequently, we used several chromosomal DNA clones to screen c-DNA libraries (kindly provided by the Hogness' laboratory) for *Antp* cDNAs. First, an embryonic and a pupal c-DNA clone were isolated. By hybridizing these cDNAs back to the chromosomal DNA from the walk, the *Antp* exons were mapped and shown to be spread over ~ 100 kb, indicating that the *Antp* gene is extremely large. Furthermore, the exon map showed that the two transcripts represented by the two clones arise by differential splicing. However, there was also cross hybridization outside of *Antp* gene as defined by deletions removing only *Antp* and leaving the flanking genes intact. "Under stringent hybridization conditions, weak homology with both the 903 and 909 cDNA probes was detected at position 190 kb". "These findings are being investigated further".[17] This was in fact the first sign of the *fushi tarazu* (*ftz*) gene, which was subsequently cloned by Atsushi Kuroiwa and also the first sign of the homeobox. I go through this history in some detail because my recollections have been challenged by McGinnis and Lawrence.[18] "...there was an anomalous band in Garber's gel that could have led to discovery. However, it was attributed to overloading of the gel, lumped into the 'uninterpretable results' category and not followed up". This is completely wrong;[19] it was followed up and shown to be reproducible by both c-DNA probes and this finding led on one hand to the isolation of ftz gene by Atsushi Kuroiwa and on the other hand to the discovery of the homeobox, since the cross-hybridization is due to the homeobox being present in both genes, *Antp* and *ftz*. In fact, since Richard Garber was leaving the laboratory around that time, I specifically asked Bill McGinnis to define and sequence the region of cross-homology. He may not remember that, but I know this for a fact. At that time, Bill McGinnis was the expert on low stringency hybridization and sequencing in the laboratory, and he certainly did a great job on the homeobox. My views have been confirmed by both Richard Garber and Atsushi Kuroiwa, and Bill McGinnis' statement "It is always possible that Gehring sensed or knew in 1982 that Garber's band was a crucial clue which should be the basis of further investigation"[19] is absolutely correct. Both Richard Garber and I expected to find cross-homology between homeotic genes, because the Lewis model assumes that these genes arose by tandem duplication. However, the finding of a precisely defined box with a remarkable degree of evolutionary conservation, came as a total surprise.

After finding that the homeobox is present in both *Antp* and *ftz* genes, I asked Pierre Spierer and David Hogness to send us the corresponding clones from *Ubx* and sure enough Bill McGinnis identified the homeobox by low stringency hybridization in the 3' exon of Ubx. Eureka at this point we knew that we had discovered something important and coined the term homeobox. We then went on to use the homeobox as a probe to screen a genomic library for the presence of the homeobox in other homeotic genes and identified two other homeotic genes, *Deformed* (*Dfd*) in the ANTP-Complex and *abdominal-A* (*abd A*) in the BX-C.[20] It took two and a half years to isolate *Antp* and by using the homeobox as a probe we were able to clone a dozen homeobox-containing genes in just a few months. The DNA sequence homology was shown to be due to a highly conserved protein-coding sequence, the homeodomain.

Independently, the ANTP-Complex had also been cloned by Scott et al[21] and later they also found the crosshomology between *Antp*, *ftz* and *Ubx*,[22] but my group immediately went on to show that the homeobox was not only found in insects, but it is also found in vertebrates, such as chickens, mice and humans,[23] and in collaboration with Eddy De Robertis the first vertebrate Hox gene from the frog Xenopus laevis was cloned[24] and shown to be identical to the *Antp* homeodomain at 55 out of 60 amino acid positions. The first mouse homeobox genes were cloned in collaboration with Frank Ruddle who happened to be on sabbatical leave in my laboratory at that time.[25] If the homeobox had been confined to insects it would have been merely an interesting curiosity, but its finding in vertebrates and later in all metazoan indicated that we had discovered a universal principle in the genetic control of animal development. Despite a bewildering diversity in their modes of development, all animals share a common set of regulatory genes involved in the determination of the body plan, in particular in the specification of the antero-posterior body axis.

Specification and Redesigning of the Body Plan

Antp loss-of-function mutants die as late embryos and show a homeotic transformation of the second thoracic segment towards the first (T2→T1), whereas the dominant gain-of-function mutants transform T1 and the posterior head segments (H) towards T2 (H, T1→T2) i.e., in the opposite direction. Similarly, in the adult fly the gain-of-function mutants exhibit a dramatic transformation of the antennae into second legs (Fig. 1), indicating that *Antp* specifies the second thoracic segment, and not the antennal segment. This is consistent with our in situ hybridization data. In parallel to the work on *Antp*, Ernst Hafen and Michael Levine developed a method for in situ hybridization to detect transcripts of a given gene on frozen sections, and these methods were immediately used to localize the *Antp* transcripts.[26] It was found that the highest concentration of transcripts is detected in the ganglion cells of T2, which is consistent with the notion that *Antp* specifies T2. Upon the development of methods for in situ hybridization to embryos in whole mount preparations, a single band of *Antp* transcripts could be detected in the T2 region as early as the blastoderm stage, when initially the body plan is laid down (Fig. 2).

Later studies have revealed that the body plan is specified in Drosophila in four conceptual steps in all of which homeobox genes are involved (Fig. 2). In a first step the maternal mRNA of the homeobox gene *bicoid* (*bcd*) is localized in the anterior pole of the oocyte as a "cap". Upon fertilization the bicoid mRNA is translated into BCD protein which forms a concentration gradient with its highest concentration at the anterior pole. BCD protein enters the nuclei to exert its gene regulatory function, whereas the cytoplasmic BCD protein binds to *caudal* (*cad*) mRNA and induces its degradation at the anteriore pole. Initially, *caudal* mRNA is uniformly distributed in the egg, but upon degradation by BCD it forms a gradient, which is in fact the first gradient that was described in the Drosophila egg.[27] Upon translation the CAD protein forms an opposite gradient to that of BCD. In a third step these gradients are subdivided into a repetitive pattern of body segments. First by gap genes which are expressed in regions comprising several segments, then by pair-rule genes which are expressed in alternate segments, and finally by segment-polarity genes that are expressed in every segment. Most of the gap genes belong to the Zinkfinger class of transcription factors, but both *orthodenticle* (*otd*) and *empty spiracles* (*ems*) that have an important evolutionary conserved role in head development contain a homeobox. Among the pair-rule class e.g., *fushi tarazu* (*ftz*) and *even-skipped* (*eve*) are homeobox genes, and among the segment polarity *genes engrailed* (*en*) was also cloned on the basis of its homeobox homology.[28] In a fourth step the identity of each individual segment is specified by the Hox genes, i.e., by those homeobox genes which are clustered in the ANT-C and BX-Complexes.

In order to prove definitively that the gene which we had cloned was indeed *Antp* we should have demonstrated that the cloned DNA segment was capable of rescuing an $Antp^-$ mutation. This proved to be very difficult since the *Antp* gene is 100 kb long, and such large DNA segments are difficult to use for transformation. Therefore, we took an alternative approach, which at the same time was directed towards understanding of the nature of the dominant gain-of-function mutants. Most *Antp* mutants are caused by chromosomal inversions with one breakpoint in the *Antp* gene and the other outside at many different chromosomal sites. The enigma was how one could split a gene in the middle and still retain its activity. However, the breakpoints of the *Antp* inversions were actually not splitting the gene in the middle, but rather they left the protein-coding region intact and fused it to another gene, potentially to another promoter or cis-regulatory region leading to an ectopic misexpression of the ANTP protein. Since it is highly unlikely that all known *Antp* inversions fuse the ANTP-coding region to an antenna-specific promoter at the other breakpoint of the inversion, we assumed that the antennal disc was probably the weak spot in the circuitry and was transformed by the ectopic expression of *Antp*, whereas other imaginal discs in which *Antp* might also have been expressed were more stably determined. These considerations led me to the idea of using a heat-shock promoter to ectopically express *Antp* all over the fly, before antennal determination had taken place, in order to induce antenna-to-leg transformations. Stephan Schneuwly, with the help of Roman Klemenz, carried out this experiment and as predicted the antennae were transformed into second legs, and the top of the head into dorsal mesothorax (Fig. 3). This was the first successful attempt at redesigning the body plan.[29] At the same time this experiment proved that we had cloned the *Antp* gene and that the dominant gain-of-function mutations were due to the ectopic expression of an at least partially active gene product.

Besides the antenna-to-leg transformation the heatshock-induced ectopic over expression induces the transformation of the dorsal head capsule into notum (dorsal T2), a structure derived from the wing disc. The $Antp^{ctx}$ allele also transforms the dorsal most part of the compound eyes into wings, but most other alleles do not. Carroll et al[30] have claimed that *Antp*

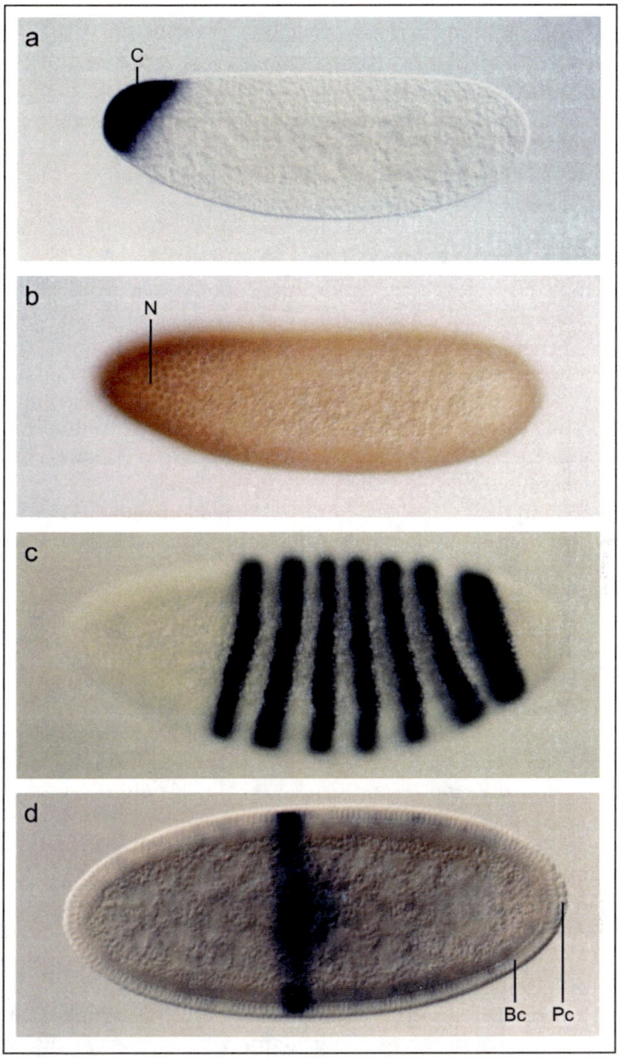

Figure 2. Four conceptual steps in the establishment of the body plan in Drosophila: a) *bicoid* mRNA localized in a cap (C); at the anterior pole of the egg; b) Bicoid protein forming an antero-posterior concentration gradient in the nuclei (N) of the early preblastoderm embryo; c) Formation of a repetitive pattern of seven stripes by the pair-rule segmentation gene fushi tarazu (in situ hybridization); d) Expression of the homeotic *Antennapedia* gene in a single stripe in the mesothoracic region of the blastoderm embryo. Bc blastoderm cells, Pc pole cells (in situ hybridization).

does not specify the wing disc since a wing disc primordium is formed in *Antp*⁻ embryos. However, they did not show that this primordium can infact develop into a wing. Our recent genetic mosaic experiments show that mosaic clones of *Antp*⁻ cells do not survive when surrounded by wild-type cells and the wings do not contain any *Antp*⁻ cell clones, whereas such clones are viable on the legs, the head and the abdomen, indicating that *Antp*⁺ is required in wing specification. The reason for the failure to induce eye-to-wing transformations by most

Antp gain-of-function mutants is due to the fact that in order to install the wing developmental program, the eye program first has to be switched off. However, switching off the eye program initiated by eyeless (*ey*) leads to an ey⁻ phenotype, i.e., cell death of the eye forming cells by apoptosis. By coexpression of *Antp*⁺ and the antiapoptotic genes p35 of baculorvirus or *Notch* we have been able to induce eye-to-wing transformations. Therefore, *Antp* specifies the entire T2 segment, including a pair of middle legs and a pair of wings. After the blastoderm stage *Antp* is also expressed more posterior in specific cells of T3 and of all abdominal segments, where its function is largely repressed by the posterior Hox genes.

The Structure and Function of the Homeodomain

The first hint at the possible function of the homeodomain came from a sequence comparison between the homeodomain and the gene regulatory proteins encoded by the yeast mating-type genes MATa1 and $\alpha 2$[31] which showed a weak, but significant similarity. This suggested that homeodomain proteins might be transcriptional regulators and that the homeodomain may represent their DNA binding domain. Sequence comparisons led to the further speculation that the homeodomain might contain a helix-turn-helix motif like the one found in prokaryotic repressors and activators.[31,32] These hypotheses were tested by analyzing the structure and function of the homeodomain in great detail. The ANTP homeodomain was expressed in bacteria, and the isolated and purified homeodomain was found to bind with high affinity ($K_D \sim 10^{-9}$ M) to specific oligonucleotides from putative binding sites.[33] Therefore, the homeodomain does represent a DNA binding domain. In contrast to prokaryotic repressors and activators, the homeodomain binds as a monomer to DNA and the consensus binding sequences in the DNA are rather loosely defined.

In a fruitful collaboration between Kurt Wüthrich's and my group, the three-dimensional structure of the Antp homeodomain polypeptide was determined at high resolution in solution by NMR spectroscopy.[34] The core structure of the homeodomain is formed by four α-helical regions. Helix 1 is connected by a loop to helix 2 which is antiparallel to helix 1, and separated by a tight turn from a well defined helix 3 which is elongated by a more flexible helix 4. Helix 2, the connecting turn, and helix 3/4 constitute the helix-turn-helix motif as first described in prokaryotic gene regulatory proteins. Eventhough there is hardly any amino acid sequence conservation, the three-dimensional structures of these motifs are essentially superimposable.

In order to analyze the homeodomain-DNA interactions we also solved the structure of the *Antp* homeodomain-DNA complex (Fig. 4) with a 14bp consensus binding site 5'-GAAAGCCATTAGAC-3'.[35,36] As expected the "recognition" helix 3/4 contacts the DNA specifically in the major groove, but the turn of the helix-turn-helix motif is shifted away from the DNA by ~7 Å as compared to prokaryotic repressor-DNA complexes. Other important contacts are established between the N-terminal arm, which is flexibly disordered insolution, and reaches into the minor groove of the DNA. The loop between helices 1 and 2 contacts the DNA backbone in the major groove. These data were subsequently confirmed by X-ray crystallography of other heomeodomain-DNA complexes.[37,38] A refined NMR analysis indicates that the invariant Asp51 and probably also Gln50 in helix 3 are in a slow dynamic equilibrium between two different cytosine residues in the DNA (see below).

Some of the essential features of the homeodomain-DNA interactions have been verified by genetic experiments in vivo. The functional analysis of different homeodomains in *Antp*, *ftz*, *paired* (*prd*) and *bicoid* (*bcd*) showed that residue 50 is of crucial importance in DNA sequence recognition.[39,40,41] The *ftz* homeodomain has a glutamine (Gln) at position 50 and, like *Antp*, it binds with high affinity to the consensus sequence CCATTA, whereas *bcd* has a lysine (Lys) at this position and preferentially binds to GGATTA.[42] As mentioned above the NMR spectroscopy shows that in the Antp homeodomain-DNA complex Gln50 contacts the two C residues adjacent to the ATTA core motif. In order to test the importance of these contacts for binding specificity in vivo, we carried out a second site suppression

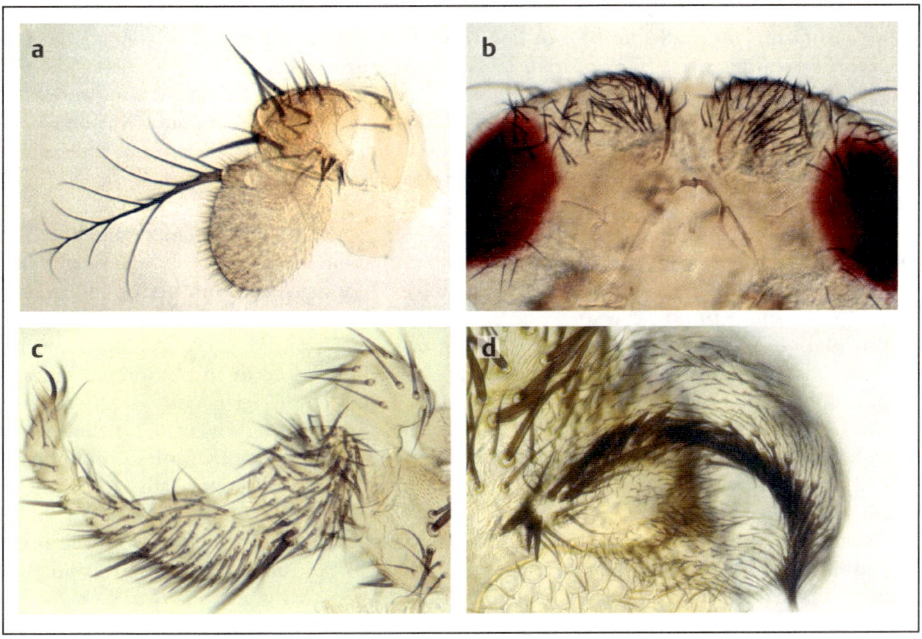

Figure 3. Redesigning of the body plan by ectopic overexpression of *Antennapedia*: a) Wildtype antenna; b-d) heat-induced ectopic overexpression of *Antennapedia* under a heat shock promoter (hs-*Antp*); b) head-to-notum transformation; c) antenna-to-second leg transformation d) eye-to-wing transformation.

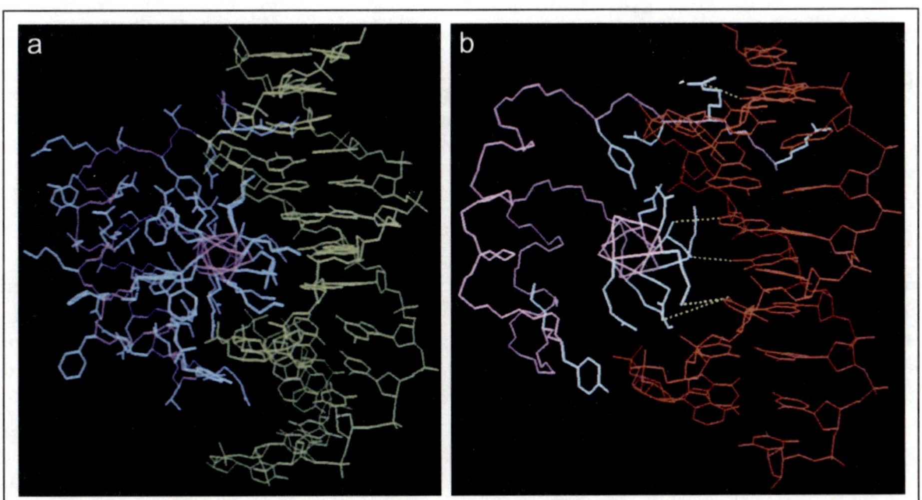

Figure 4. Solution structure of the Antennapedia homeodomain-DNA complex as determined by NMR spectroscopy a) Double-stranded DNA in yellow; homeodomain backbone in magenta, homeodomain side chains in blue b) DNA in red; homeodomain backbone in magenta, amino acid side chains contacting the DNA in blue. Amino acids contacting specific bases are connected by dotted yellow lines.

experiment using the ftz homeodomain which binds to its own autoregulatory enhancer.[41] Upon mutating the multiple ftz binding-sites in the enhancer to GGATTA i.e., to bcd concensus binding sites, FTZ protein binding and autoregulation is largely abolished. A single amino acid substitution from Gln50 to Lys50 in the ftz transgene suppresses this mutant effect and restores the normal pattern of seven stripes. Thus, the substitution of two bases in the binding site can be overcome by substituting a single amino acid in the homeodomain. This demonstrates a direct interaction of amino acid 50 in the homeodomain and the two CC (or GG) residues in the binding site.

The functional importance of the flexible N-terminal arm of the homeodomain which binds in the minor groove was analyzed by comparing the functional specificities of the *Antp* and *Sexcombs reduced* (Scr) which are very similar. By exchanging only four specific amino acids in the N-terminal arm, the specificity of a hybrid protein can be changed from that of *Scr* to that of *Antp* capable of inducing antenna-to-leg transformations.[43] Two of these Scr-specific amino acids at position 6 and 7 are threonine and serine and occur in phosphorylated and nonphosphorylated form. As shown by substituting T and S by either AA or two DD residues mimicking the dephorylated and the phosphorylated form respectively, we have shown that the phosphorylated form is inactive. Dephosphyorylation of Tand S by protein phosphatase PP2A activates the Scr homeodomain.[44] The inactivity of the phosphorylated form can be attributed to impaired DNA binding.

However, specific DNA binding is not the only function of the homeodomain. In addition the homeodomain of *bicoid* (*bcd*) is capable of binding *caudal* (*cad*) mRNA in vitro, and exert translational control of *cad* mRNA through bcd-binding in vivo.[45]

Furthermore, our analysis of repression of eye development by ectopically expressed ANTP protein has shown that ANTP and Eyeless (EY) proteins interact at the posttranslational level[46] and that repression occurs via protein-protein interactions due to the binding of the Antp homeodomain to the paired domain and the homeodomain of *eyeless* (*ey*).

Finally, the homeodomain contains an oligopeptide sequence called penetratin, which allows it to penetrate through the plasma membrane.[47] For the outgrowing axons from Xenopus retinal neurons the engrailed protein can serve as either an attractant or repellent in axon guidance.[48]

Evolution of Homeobox Genes

Homeoboxes have been found in fungi, plants and animals. In each "kingdom" homeobox genes occupy a key position in the genetic control of either cell differention, morphogenesis and or body plan specification. In fungi which exhibit only a limited capacity for cell differentiation homeobox genes control the mating-types in both budding and in fission yeasts, and in the determination of filamentous versus yeast-like growth forms in *Ustilago*.[49] In plants they play a different role than in animals. The "homeotic" genes which specify the various wirls of the flower encode MADS box proteins; however, the knotted homeobox-containing gene in maize, for example[50] has an important function in the meristem and therefore, in morphogenesis. In animals homeobox genes can be traced back to the most primitive metazoan, to cnidarians and sponges. Only limited data are available on protozoa, but the recently available genome sequence of *Dictyostelium* which is in the transition zone between uni- and multicellular organisms, reveals a substantial number of homeobox genes.

The degree of sequence conservation of the homeodomain is extremely high indicating strong functional constraints leading to a highly selective pressure to retain the homeobox sequences constant. For example, the human Hox gene which is most closely related to the *Antp* gene of Drosophila, shares 59 out of 60 amino acids with Drosophila and 60 out of 60 with that of the honey bee. However, it is not only the amino acid sequence which is so highly conserved, but also the colinear arrangement of the Hox genes.

The Hox genes of Drosophila are arranged in two clusters, the ANT-C and the BX-C, but this is a relatively recent arrangement, since a single cluster is found in more primitive insects as e.g., in the beetle *Tribolium*, and in *Drosophila virilis* the breakpoint separating the two clusters is between *Ubx* and *abdA* rather than between *Antp* and *Ubx* as it is in *Drosophila melanogaster*. This indicates that the splitting into two clusters is a relatively recent evolutionary event which has occurred several times. In vertebrate evolution the Hox clusters was duplicated at least twice to give rise to 4 Hox clusters in mice and humans in all of which the colinear arrangement of the paralogous genes has been conserved. A single, but complete cluster is found in *Amphioxus*, the most primitive known chordate ancestor of vertebrates. A single compact cluster of Hox genes has also been found in the nemertean *Lineus*, which may represent one of the most primitive bilaterian animals still existing.[51] Apparently, Lineus has evolved rather slowly, since in fast evolving organisms like Drosophila, and even more in the nematode *C. elegans* the Hox cluster has been split up. In contrast to *Amphioxus*, the nemerteans are nonsegmented animals indicating that the Hox-cluster formed prior to the evolution of body segments.

A hypothesis on the evolution of the Hox gene clusters which is based on unequal crossing over giving rise to serial tandem duplication followed by functional divergence has been put forward.[52] This model is compatible with the phylogenetic tree reconstructed by Nei.[53] However, it is difficult to obtain more direct evidence for Hox gene evolution, since most of the early evolutionary stages have been lost, and even candidates for the urbilaterian ancestor like the nemerteans have a rather complete Hox cluster.

The question of whether the vertebrate Hox genes are also functionally homologous to the homeotic genes of Drosophila has been tackled in the mouse by constructing both gain- and loss-of-function mutants. Along the same lines as our Antp gain-of-function experiment,[54] Kessel and Gruss have overexpressed Hoxa-7 in transgenic mice and induced an extracervical vertebra, a proatlas.[54] Mammals have generally 7 cervical vertebrae whereas reptiles have a much more variable number. In the transgenic mutant mice the basioccipital bone is transformed into a proatlas as it is found in some reptiles. This is a transformation in the posterior direction. By contrast, targeted disruption of Hoxd-3 leads to an anterior transformation of the first and second cervical vertebrae, the atlas and the axis,[55] in much the same way as in Drosophila.

The most direct way of testing the mouse Hox genes for their putative homeotic function is to introduce the mouse genes into Drosophila and over express them ectopically (i.e., ubiquitously) under a heat-shock promoter. Heat-induced over expression of Hoxb-6, an *Antp* homolog, leads to the transformation of the antenna into leg structures,[56] whereas over expression of Hoxa-5, the *Scr* homolog, leads to a transformation of the T2 and T3 segments of the transgenic larvae into T1, the same transformation that is caused by the Drosophila *Scr* gene.[57]

However, some of the genes in the Hox complexes of Drosophila have evolved considerably and acquired new functions. This is particularly the case of the genes bicoid (*bcd*), *zerknüllt* (*zen*) and *fushi tarazu* (*ftz*), who are important for the establishment of the body plan in the early embryo and do not have the same function in vertebrates. During insect evolution the Hox genes *zen* and *ftz* evolved new developmental roles in dorso-ventral axis specification and segmentation respectively, and lost their ancestral roles in specifying the antero-posterior axis. In diptera the *zen* gene has been duplicated and gave rise to *bcd*, which has an essential role in the establishment of antero-posterior polarity of the egg, but is not found in other insects like the beetles.[58]

Although some of the Hox genes have acquired new functional roles during evolution, the major driving force for morphological evolution is not based on the evolution of new genes, or point mutations in preexisting genes, but rather on the evolution of cis-regulatory regions involved in the control of gene expression.[59] This view differs radically from the hypotheses based upon classical population genetics which have to be revised.

References

1. Morgan TH. Embryology and Genetics. New York: Columbia University Press, 1934.
2. Bridges C, Morgan TH. Carnegie Inst Wash Publ 1923; 317:1-251.
3. Lewis EB. Genes and developmental pathways. Am Zool 1963; 3:33-56.
4. Lewis EB. A gene complex controlling segmentation in Drosophila. Nature 1978; 276:565-70.
5. Garcia-Bellido A, Ripoll P, Morata G. Developmental compartmentalisation of the wing disk of Drosophila. Nature New Biol 1973; 245:251-253.
6. Sánchez-Herrero E, Vernós RM, Morata G. Genetic organization of Drosophila bithorax complex. Nature 1985; 313:108-113.
7. Lewis EB. Clusters of master control genes regulate the development of higher organisms. The Journal of the American Medical Association (JAMA) 1992; 267:1524-1531.
8. Ronshaugen M, Biemar F, Piel J et al. The Drosophila microRNA iab-4 causes a dominant homeotic transformation of halteres to wing. Genes and Dev 2005; 19:2947-2952.
9. Bender W, Spierer P, Hogness DS. Chromosomal walking and jumping to isolate DNA from the Ace and rosy loci and the bithorax complex in Drosophila melanogaster. J Mol Biol 1983; 168:17-33.
10. Bender W, Akam M, Karch F et al. Molecular genetics of the bithorax complex in Drosophila melanogaster. Science 1983; 221:23-29.
11. Karch F, Weiffenbach B, Peifer M et al. The abdominal region of the bithorax complex. Cell 1985; 43:81-96.
12. Aravin AA, Lagos-Quintana M, Yalcin A et al. The small RNA profile during Drosophila melanogaster development. Dev Cell 2003; 5:337-350.
13. Gehring W. Bildung eines vollständigen Mittelbeines mit Sternopleura in der Antennenregion bei der Mutante Nasobemia (Ns) von Drosophila melanogaster. Jul Klaus Arch 1966; 41:44-54.
14. Le Calvez J. Mutation ®stapediax" hétérozygote dominate, homozygote léthale chez Drosophila melanogaster. Bull Biol France et Belgique 1948; 82:97-113.
15. Lewis E. Report on the mutants Antennapedia-Bacon and Antennapedia-Yu. Drosoph Inf Serv 1956; 30:76.
16. Jorgensen EM, Garber RL. Function and misfunction of the two promoters of the Drosophila Antennapedia gene. Genes and Dev 1987; 1:544-555.
17. Garber RL, Kuroiwa A, Gehring WJ. Genomic and cDNA clones of the homeotic locus Antennapedia in Drosophila. EMBO J 1983; 2:2027-2036.
18. McGinnis W, Lawrence PA. Historical Transformations. Nature 1999; 398:301-302.
19. Gehring W. Lifting the lid on the homeobox discovery. Nature 1999; 399:521-522.
20. McGinnis W, Levine MS, Hafen E et al. A conserved DNA sequence in homeotic genes of the Drosophila Antennapedia and bithorax complex. Nature 1984; 308:428-433.
21. Scott MP, Weiner AJ, Hazelrigg TI et al. The molecular organization of the Antennapedia Locus of Drosophila. Cell 1983; 35:763-776.
22. Scott M, Weiner A. Structural relationships among genes that control development: Sequence homology between Antennapedia, Ultrabithorax, and fushi tarazu loci of Drosophila. Proc Natl Acad Sci USA 1984; 81:4115-4119.
23. McGinnis W, Garber RL, Wirz J et al. A homologous protein-coding sequence in Drosophila homeotic genes and its conservation in other metazoans. Cell 1984; 37:403-408.
24. Carrasco AE, McGinnis W, Gehring WJ et al. Cloning of an X.laevis gene expressed during early embryogenesis that codes for a peptide region homologous to Drosophila homeotic genes. Cell 1984; 37:409-414.
25. McGinnes W, Hart CP, Gehring WJ et al. Molecular cloning and chromosome mapping of a mouse DNA sequence homologous to homeotic genes of Drosophila. Cell 1984; 38:675-680.
26. Levine M, Hafen E, Garber RL et al. Spatial distribution of Antennapedia transcripts during Drosophila development. EMBO J 1983; 2:2037-2046.
27. Mlodzik M, Fjose A, Gehring WJ. Isolation of caudal, a Drosophila homeobox-containing gene with maternal expression, whose transcripts form a concentration gradient at the preblastoderm stage. EMBO J 1985; 4:2961-2969.
28. Fjose A, McGinnis WJ, Gehring WJ. Isolation of a homeobox-containing gene from the engrailed region of Drosophila and the spatial distribution of its transcripts. Nature 1985; 313:284-289.
29. Schneuwly S, Klemenz R, Gehring WJ. Redesigning the body plan of Drosophila by ectopic expression of the homeotic gene Antennapedia. Nature 1987; 325:816-818.
30. Carroll SB, Weatherbee SD, Langeland JA. Homeotic genes and the regulation and evolution of insect wing number. Nature 1995; 375:58-61.
31. Shepherd JCW, McGinnis W, Carrasco AE et al. Fly and frog homeodomains show homologies with yeast mating type regulatory proteins. Nature 1984; 310:70-71.

32. Laughon A, Scott MP. Sequence of a Drosophila segmentation gene: Protein structure homology with DNA-binding proteins. Nature 1984; 310:25-31.
33. Affolter M, Percival-Smith A, Müller M et al. DNA binding properties of the purified Antennapedia homeodomain. Proc Natl Acad Sci USA 1990; 87:4093-4097.
34. Qian YO, Billeter M, Otting G et al. The structure of the Antennapedia homeodomain determined by NMR spectroscopy in solution. Comparison with prokaryotic repressors. Cell 1989; 59:573-580.
35. Otting G, Qian YQ, Billeter M et al. Protein-DNA contacts in the structure of a homeodomain-DNA complex determined by nuclear magnetic resonance spectroscopy in solution. EMBO J 1990; 9:3085-3092.
36. Billeter M, Qian YQ, Otting G et al. Determination of the nuclear magnetic resonance solution structure of an Antennapedia homeodomain-DNA complex. J Mol Biol 1993; 234:1084-1093.
37. Kissinger CR, Liu B, Martin-Blanco E et al. Crystal structure of an engrailed homeodomain-DNA complex at 2.8 Å resolution: A framework for understanding homeodomain-DNA interactions. Cell 1990; 63:579-590.
38. Wolberger C, Vershon AK, Liu B et al. Crystal structure of a MAT α2 homeodomain-operator complex suggests a general model for homeodomain-DANN interactions. Cell 1991; 67:517-528.
39. Hanes SD, Brent R. DNA specificity of the bicoid activator protein is determined by homeodomain recognition helix residue 9. Cell 1989; 57:1275-83.
40. Treisman J, Gönczy P, Vashishtha M et al. A single amino acid can determine the DNA binding specificity of homeodomain proteins. Cell 1989; 59:553-562.
41. Schier AF, Gehring WJ. Direct homeodomain-DNA interaction in the autoregulation of the fushi tarazu gene. Nature 1992; 356:804-807.
42. Percival-Smith A, Müller M, Affolter M et al. The interaction with DNA of wild-type and mutant fushi tarazu homeodomains. EMBO J 1990; 9:3967-3974.
43. Furukubo-Tokunaga K, Flister S, Gehring WJ. Functional specificity of the Antennapedia homeodomain. Proc Natl Acad Sci USA 1993; 90:6360-6364.
44. Berry M, Gehring WJ. Phosphorylation status of the SCR homeodomain determines its functional activity: Essential role for protein phosphatase 2A,B'. EMBO J 2000; 19:2946-2957.
45. Rivera-Pomar R, Niessing D, Schmitt-Ott U et al. RNA binding and translational suppression by bicoid. Nature 1996; 379:747-749.
46. Plaza S, Prince F, Jäger J et al. Molecular basis for the inhibition of Drosophila eye development by Antennapedia. EMBO J 2001; 20:802-811.
47. Joliot A, Prochiantz A. Transduction peptides: From technology to physiology. Nat Cell Biol 2004; 6:189-196.
48. Brunet I, Weinl C, Piper M et al. The transcription factor Engrailed-2 guides retinal axons. Nature 2005; 438:94-98.
49. Schulz B, Banuett F, Dahl M et al. The b alleles of U. maydis, whose combinations program pathogenic development, code for polypeptides containing a homeodomain-related motif. Cell 1990; 60:295-306.
50. Vollbrecht E, Veit B, Sinha N et al. The developmental gene Knotted-1 is a member of a maize homeobox gene family. Nature 1991; 350:241-243.
51. Kmita-Cunisse M, Loosli B, Bièrne J et al. Homeobox genes in the ribbonworm Lineus sanguineus: Evolutionary implications. Proc Natl Acad Sci USA 1998; 95:3030-3035.
52. Gehring WJ. Master control genes in development and evolution: The homeobox story. New Haven: Yale University Press, 1998.
53. Zhang J, Nei M. Evolution of antennapedia-class homeobox genes. Genetics 1996; 142:295-303.
54. Kessel M, Gruss P. Homeotic transformation of murine vertebrae and concomitant alteration of Hox codes induced by retinoic acid. Cell 1991; 67:89-104.
55. Condie B, Capecchi MR. Mice homozygous for a targeted disruption of Hoxd-3 (Hox-4.1) exhibit anterior transformations of the first and second cervical vertebrae, the atlas and the axis. Development 1993; 119:579-595.
56. Malicki J, Schughart K, McGinnis W. Mouse Hox-2.2 specifies thoracic segmental identity in Drosophila embryos and larvae. Cell 1990; 63:961-967.
57. Zhao JJ, Lazzarini RA, Pick L. The mouse Hox-1.3 gene is functionally equivalent to the Drosophila Sex combs reduced gene. Genes Dev 1993; 7:343-354.
58. Mann RS, Carroll SB. Molecular mechanisms of selector gene function and evolution. Curr Opin Gent Dev 2002; 12:592-600.
59. Carroll SB, Grenier JK, Weatherbee SD. From DNA to diversity. Molecular Genetics and the Evolution of Animal Design, 2nd ed. Malden: Blackwell Scientific, 2005.

Chapter 2

Expression of *Hox* Genes in the Nervous System of Vertebrates

Christof Nolte and Robb Krumlauf*

Abstract

The vertebrate nervous system is a major site of *Hox* gene expression and function. Studies on the patterns of expression, regulation and function of the vertebrate *Hox* gene family have played a key role in aiding our understanding of the basic ground plan of the CNS and processes that control how unique regional character is established and maintained in this complex organ system. This chapter will document the nature of the ordered patterns of *Hox* expression and link them with their regulation and functional roles in the nervous system.

Introduction

The adult vertebrate nervous system is a complex organ formed by a progressive series of intricately coordinated events that generate the components and circuitry essential for its function. While there is considerable diversity in the brain and CNS associated with higher order functions in vertebrates, underlying this final complexity there is a highly conserved basic genetic program that governs the fundamental patterns by which the central nervous system (CNS) is formed.

A central question is how this highly organized system arises in the body? For example, what 'instructs' a developing neuron to form in a particular location, correctly project to its targets or interact with other neurons? Part of the answer to these questions involves the products of the *Hox* transcription factor gene family that during development specify anterior-posterior (A-P) identity in different tissues, including the CNS. The purpose of this review is to highlight the expression, regulation and roles of *Hox* genes in patterning CNS development. A major focus will be placed on the hindbrain because of the considerable number of studies in different vertebrate models that demonstrate the important conserved role the *Hox* genes play in regulating the basic ground plan of this region of the CNS.

The *Hox* Gene Family

Initially identified and characterized in *Drosophila* as the genes associated with homeotic transformation, *Hox* genes were subsequently identified in vertebrates by virtue of their highly conserved homeobox sequences.[1-5] From a concerted effort of many different laboratories, a total of 39 *Hox* genes have been identified in most vertebrates, except the ray-finned fishes. This large gene family is organized into four separate chromosomal clusters that can contain up to 11 genes with each complex (Fig. 1).[5,6] Based on similarities between the putative translated

*Corresponding Author: Robb Krumlauf—Stowers Institute for Medical Research, 1000 50th Street, Kansas City, Missouri 64110, U.S.A. Email: rek@stowers-institute.org

HOX Gene Expression, edited by Spyros Papageorgiou. ©2007 Landes Bioscience and Springer Science+Business Media.

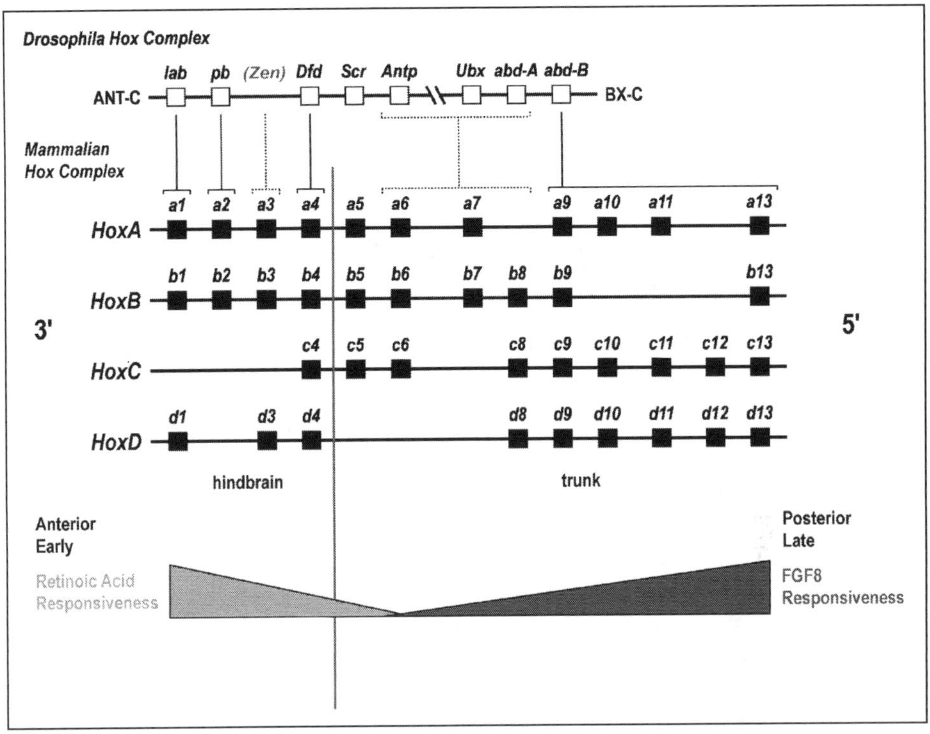

Figure 1. The *Drosophila* and vertebrate *Hox* complexes. The *Drosophila Hox* Complex is split into two complexes: the *Antennapedia* Complex (ANT-C) and the *Bithorax* Complex (BX-C). In general, the vertebrate *Hox* genes are organized into four clusters found on separate chromosomes. However, in the case of zebrafish seven clusters have been reported (not shown; ref. 9). Based on protein sequence comparisons, the *Hox* genes can be organized into 1 of 13 paralogous groups. All are organized in the same transcriptional (5' to 3') direction within each complex. Most of the first four vertebrate paralogs are expressed in the hindbrain, whereas the rest are expressed in the trunk. The early 3' located *Hox* genes display the highest level of retinoic acid (RA) responsiveness, whereas the 5' members are fibroblast growth factor (FGF) responsive.

sequences of the vertebrate *Hox* genes and the *Drosophila* Hox proteins, the *Hox* complex appear to have arisen by duplication and divergence from a common ancestor. The genes can be assigned into one of thirteen paralogous groups (Fig. 1). Thus, members of the *labial* (*lab*) subfamily, group 1 *Hox* genes, include *Hoxa1*, *Hoxb1*, and *Hoxd1*. Even though insect and vertebrates shared a common ancestor over hundreds of millions of year ago, their linear organization on the chromosome is maintain between complexes as well as conserved with the order found in *Drosophila*.[7,8] However, unlike the *Drosophila Hox* complex, the vertebrate complexes are small (~120 kb) and devoid of non*Hox* genes. In ray-finned fishes there has been a subsequent round of genome-wide duplication, generating up to 8 *Hox* clusters and many more *Hox* genes.[9-12]

Regardless of the species, one of the most fascinating features of *Hox* genes is the phenomenon known as *colinearity*.[8,13-17] This term describes how the order in which each gene becomes activated matches their linear order along the chromosomal cluster. They are all transcribed in the same 5' to 3' orientation, with the 3' most genes expressed the earliest and with the most anterior borders of expression (Fig. 3). This process continues for each successive gene along the complex from the 3' to the 5' end. Consequently, the anterior borders of consecutively expressed genes are set more posteriorly than those of the earlier expressed genes, generating a nested and overlapping series of expression patterns. As a result of colinearity, unique combinations of Hox proteins are produced at different positions along the A-P axis of the developing embryo. This observation led to the proposal of a *Hox Code* whereby different combinations of Hox proteins generated by the nested domains of expression determine a cell's A-P placement and regional character within the developing hindbrain,[16,18,19] limb,[20-22] genitalia,[23] and somites of the embryonic trunk.[24,25]

Hox proteins function as transcription factors to regulate other genes and hence play a pivotal role in the pathways that specify tissue identity. Each contains a homeodomain, translated from the sequence of the homeobox, which bears high amino acid sequence similarity to the homeodomain of the archetypical *Hox* member, *Antennapedia*. The homeodomain makes major groove contacts, via a helix-turn-helix motif, and minor groove contacts, via the N-terminal arm of the homeodomain, with DNA.[26] From numerous loss-of-function and gain-of-function experiments in diverse vertebrate and invertebrate species, it has been shown that the anterior border of a *Hox* gene expression domain is important as perturbations of this border typically result in homeotic transformations and malformations.[27,28] Here we review where these borders are set in the developing hindbrain of the CNS, what transcription factors and signaling pathways initiate the setting of these borders, and what are the consequences to CNS patterning when these borders are altered.

The Embryonic Vertebrate Nervous System

The central nervous system (CNS) begins as a thickened sheet of epithelial cells that forms the neural plate. Starting at the anterior end of the embryo, the lateral edges of the neural plate will roll upwards and fuse into the neural tube, while at the posterior end of the embryo the A-P axis is extended by cell proliferation, thereby elongating the trunk and neural plate. As the cells leave this proliferative region their A-P identity or regional character becomes determined. This process occurs in both the neural tube and the adjacent paraxial mesoderm as it differentiates into somites.

Since the development of the vertebrate embryo proceeds temporally in an anterior to posterior fashion, the rostral end of the embryo is more developmentally advanced than the caudal end. Following neural induction the neural tube becomes subdivided into morphologically distinct domains: the forebrain, midbrain, hindbrain, and spinal cord. Current models are consistent with the idea that the initial state of the neural tube is forebrain-like and that the more posterior regions (midbrain, hindbrain and spinal cord) arise through a progressive transformation to more posterior fates. This is referred to as the "activation and transformation" model of neural development.[29-31] It is believed that distinct signaling pathways and processes govern each of these territories rather than a single common mechanism.[32]

The spinal cord can be further subdivided, according to the vertebra that will form to enclose it, into the cervical, thoracic, lumbar, and sacral domains. Within the cervical and thoracic domains, a third level of division can be identified as the brachial domain which spans vertebrae C5 to T1. These designations are important because different "columns" of MNs are generated in each domain. Hence in the brachial domain, the spinal cord will generate lateral column (LMC) neurons, whereas the remainder of the thoracic region will generate autonomic MNs.[a] The LMC neurons will send axons into the limb, whereas the autonomic MNs will project axons to the body wall muscles.[33]

Within a transverse cross section of the spinal cord, there are differences in the dorsal-ventral (D-V) axis as different cell layers can be distinguished (Fig. 2A,B). The inner most layer, which lines the lumen of the neural canal, is the ventricular layer. This is composed of undifferentiated, proliferating cells. It is surrounding by the gray-colored mantle layer which contains

[a] In the chick the autonomic MNs are called the Column of Terni neurons.

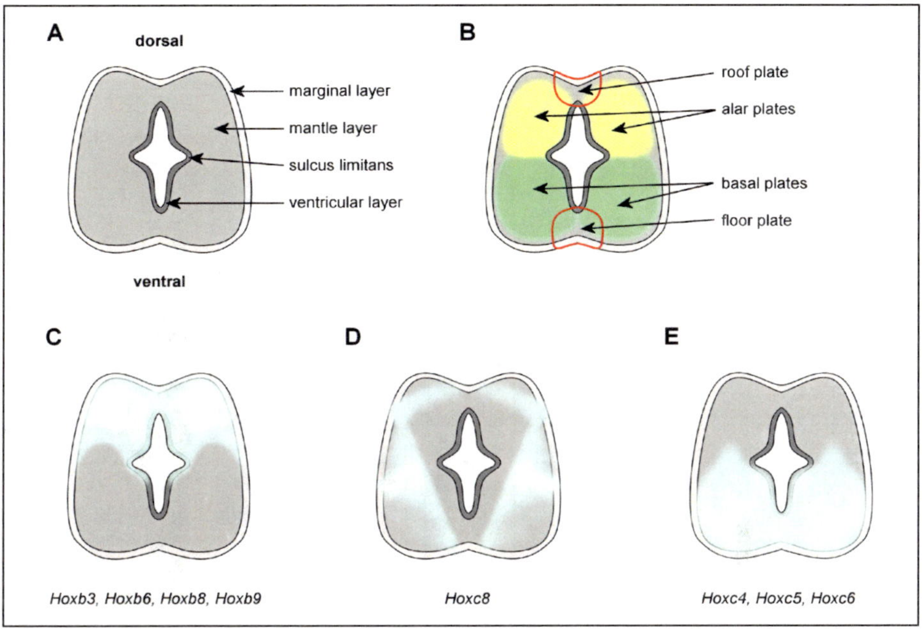

Figure 2. *Hox* gene expression in the 12.5 dpc dorsal-ventral (D-V) axis of the mouse spinal column. Different regions exist within the D-V aspect of the spinal column. A) There are three main layers in the spinal column: the inner ventricular layer, the middle mantle layer, and the outer marginal layer. B) Superimposed onto these layers, are four different "plates": the roof plate, the alar plates, the basal plates and the floor plate. The majority of *HoxB* expression occurs in the dorsal half of the spinal column predominately throughout the alar plates (C), whereas the majority of *HoxC* expression occurs in the ventral half of the spinal column throughout the basal plates (E). Note, however that *Hoxc8* is expressed in a very distinctive pattern that is not restricted to any specific plate or layer (D).

differentiating neurons. The mantle layer is enclosed by the white-colored marginal layer that contains the nerve fibers. The mantle layer can be further subdivided into the dorsal alar and ventral basal plates. The alar plate forms the sensory area while the basal plate forms the motor area of the spinal cord.

So far we have described all of the major components of the CNS, whereas the peripheral nervous system (PNS) will arise from the neural crest. The neural crest is a transitory population of cells that delaminate from the border area located between the surface ectoderm and the neural plate as the neural tube is formed. After their induction, they undergo an epithelial-to-mesenchymal transition and migrate to give rise to numerous derivatives.[34] In the head, the 'first wave' of neural crest migrate into the branchial arches while a 'second wave' migrates a short distance from the neural tube to form neural derivatives.

The brain stem is a relatively small region just anterior to the spinal cord, yet its functional significance is very great. The brain stem includes the midbrain and the hindbrain. The *Hox* genes are central to events involved in specifying the hindbrain territory[35] and furthermore they seem to be part of a highly conserved mechanism for regulating properties of this territory in vertebrates. During development, the vertebrate hindbrain undergoes a segmentation process which subdivides the neuroepithelium into a number of compartments or rhombomeres, along the anteroposterior (AP) axis.[32,36,37] Early cellular partitioning of the hindbrain into segments represents an important mechanism by which neuronal organization and diversification are

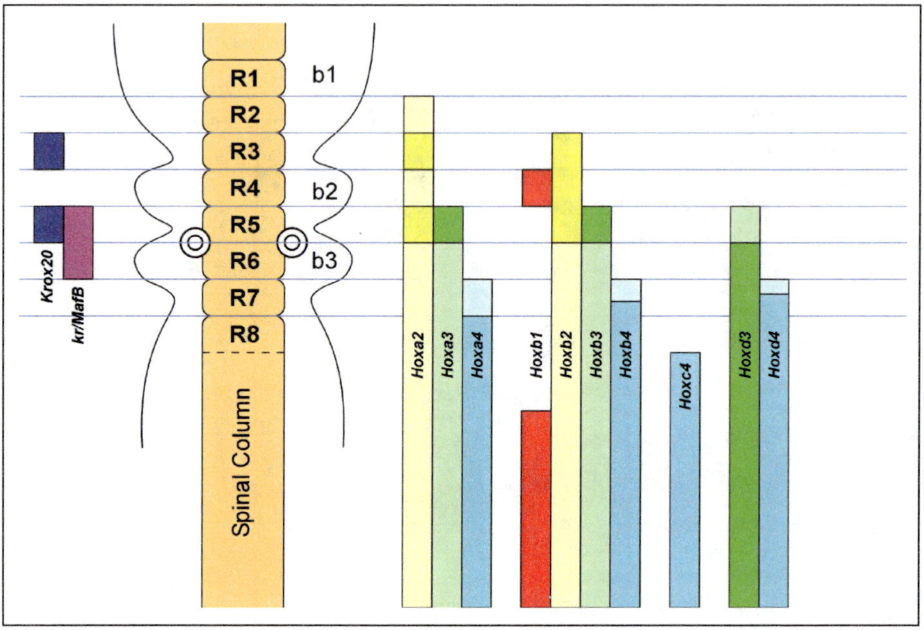

Figure 3. *Hox* gene expression in the 9.5 dpc mouse hindbrain. The vertebrate hindbrain undergoes a period of transient segmentation into 7 or 8 rhombomeres (labeled as R1 to R8). Different cranial nerves develop from the rhombomeres into the branchial arches (b1, b2, and b3). To simplify the image, these projections as well as the migrating cranial neural crest cells have not been illustrated. The position of r5 and r6 is established by the presence of the highly visible otic vesicles which develop adjacent to the hindbrain and will form the inner ear (displayed as circles flanking r5 and r6). *Krox20* and *kreisler* (*kr/MafB*) are expressed in specific rhombomeres and control the expression of *Hox* genes expressed in these segments. Ten of the 3' most located *Hox* genes are expressed in the hindbrain and their anterior borders correspond to rhombomeric boundaries. *Hoxd1* and *Hoxc4* are not expressed in the hindbrain at this stage. Initially the *Hoxb1* expression domain is continuous throughout the neural tube, but regulatory events by 9.5 dpc limit its expression to r4 within the hindbrain. For many *Hox* genes the intensity of expression through the CNS is dynamic. Lower areas of expression are depicted by lighter, colored bars.

initially established. The hindbrain is densely packed with vital structures. It contains the nuclei and fibers of the cranial nerves which innervate the muscles of the head and neck, transmit sensory information on hearing, balance and taste and control the cardiovascular and gastrointestinal systems. In addition, it contains the sensory tracts ascending from the spinal cord to the thalamus and cortex and the motor pathways descending from the forebrain to the brain stem and spinal cord. The reticular formation of the rhombencephalon contains higher order relay centers that control respiration and blood pressure as well as centers that mediate arousal and wakefulness. The distinct neuronal groups in the hindbrain are also the major source of noradrenergic and serotonergic inputs to most parts of the brain. Brain stem patterning appears to use common cellular and molecular mechanisms to generate these features because they are a shared feature of vertebrates. To understand how patterns of neuronal connectivity are established and maintained, we first need to know how a given neuronal precursor comes to occupy its particular position in the nervous system, how it acquires a specific identity appropriate to that position, and how this translates into its specific differentiated properties.

Overt and Covert Segmentation of the Nervous System

In *Drosophila, Hox* gene expression is associated with the visible segmentation of the embryo.[38] Misexpression and mutational analysis of fly *Hox* genes result in the affected segments acquiring another segment's identity (homeotic transformation).[5] One obvious example of segmentation in the vertebrate embryo is the formation of the somites from the paraxial mesoderm that will give rise to the axial skeleton, in addition to most of the musculature of the trunk. Early experiments with vertebrate *Hox* genes showed that their role in segmental identity was conserved when their loss or misexpression resulted in the transformation, and in some cases malformations, of elements of the axial skeleton.[24,25,39-47]

A similar level of serial segmentation is seen in the hindbrain which visibly subdivides into a series of seven to eight compartments called *rhombomeres* (r).[32,36,37] Morphologically these compartments are short-lived, but each acts as a separate entity displaying very little cell mixing between adjacent compartments.[37,48] Even though each rhombomere will generate a similar pool of neurons, their number and axonal projections are rhombomere specific.[49,50] Patterns of expression in combination with extensive gain- and loss-of-function analyses in different vertebrates have revealed that the distinct cellular and molecular characteristics of rhombomeres are the result of different combinations of *Hox* genes being expressed in different rhombomeres; the so-called *Hox Code.*[16,19,51-70]

Superficially the developing spinal column appears to lack the ordered pattern of segmentation that is displayed by the adjacent somites or by the hindbrain. Work with the chick neural tube suggests that it indeed lacks any form of direct segmentation[71,72] but that its local A-P character is probably determined by the adjacent somites hence there is some evidence for indirect segmentation through this route.[73-75] However, others argue that the symmetrical pattern of axonal projections along its length suggest that it displays some level of intrinsic segmentation. Regardless of the arguments of whether or not it is molecularly segmented, the adjacent somites as well as their sclerotome derivatives (the prevertebrae) are often been used as landmarks to position the anterior boundaries of *Hox* gene expression in the developing spinal column (Fig. 4).

Hox Gene Expression in the CNS

Hox *Genes from the Groups 1 to 4 Are Expressed in the Hindbrain*

As a result of colinearity *Hox* genes expressed in the hindbrain are from paralog groups 1-4, where 10 of the 12 members display segmentally-restricted domains of expression (Fig. 3).[13,19] Members from groups 5 to 13 have anterior boundaries of expression which map in the spinal cord (Fig. 4).[35] Although earlier work documented individual expression patterns of vertebrate *Hox* genes as they were initially cloned (for example; *Hoxb5,*[76] *Hoxa1,*[77,78] *Hoxb1-Hoxb3,*[16] Hunt and coworkers (1991) systematically cataloged the expression patterns of *Hox* genes that are expressed in the developing hindbrain, branchial arches, and cranial neural crest.[19] This work highlighted several important observations. First, genes within the same paralogous group are generally expressed at the same rhombomere boundary although relative levels of expression in each segment vary between paralogs. Thus, members of the 1st, 3rd, and 4th groups have anterior boundaries that coincide with the r3/r4, r4/r5, and r6/r7 boundaries, respectively (Fig. 3). Second, the boundaries between the genes in the groups 1 to 4 occur with a two-rhombomere periodicity. Thus, the anterior border of early *Hoxb1* expression is positioned two rhombomere lengths anterior to the border of *Hoxb3*, and four rhombomere lengths anterior to the border of *Hoxb4*.

Of course, there are some exceptions to these observations. First, *Hoxa2* and *Hoxb2* have different boundaries, such that *Hoxa2* is expressed up to the r1/r2 boundary[79] while *Hoxb2* is expressed up to the r2/r3 boundary.[16,19] Hence, *Hoxa2* is the only *Hox* gene expressed in r2. Second, the early pattern of *Hoxb1* later becomes restricted to r4, while *Hoxa1* is no longer expressed in the hindbrain.[16,77,78,80] Thirdly the anterior boundary of *Hoxc4* expression starts

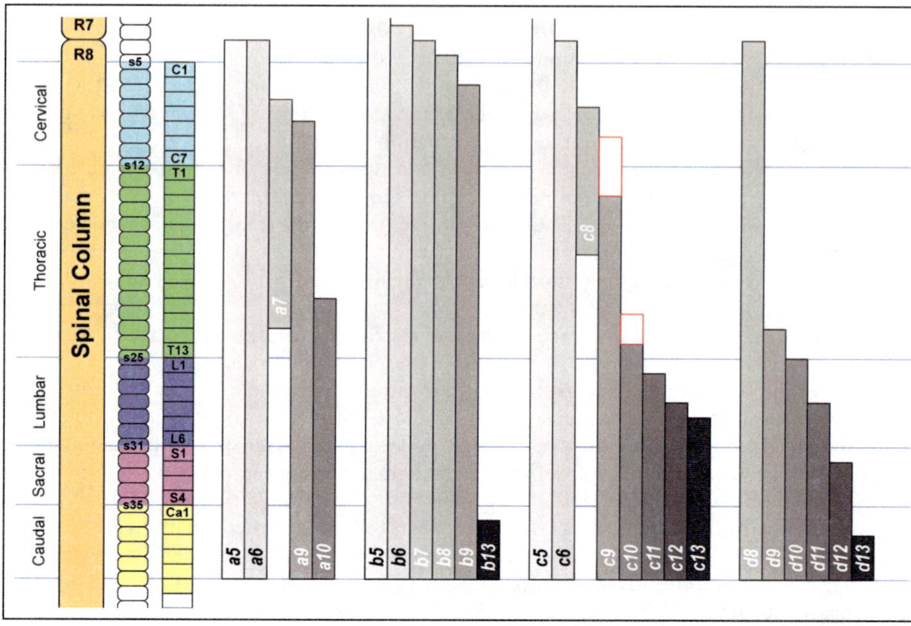

Figure 4. *Hox* gene expression in the 12.5 dpc anterior-posterior (A-P) axis of the mouse spinal column. *Hox* genes from groups 5 to 13 begin their expression in the spinal column. Since the spinal column lacks overt segmentation, the adjacent somites (s) or prevertebrae (shown as their final vertebrae segments) are used to describe *Hox* gene expression in the neural tube. In the mouse the prevertebrae will become seven cervical (C1 to C7), thirteen thoracic (T1 to T13), six lumbar (L1 to L6), four sacral (S1 to S4) and five or more caudal (Ca1 to end) vertebrae. Many of the *Hox* genes have initial anterior borders that are posterior to their borders seen at 12.5 dpc. By this stage, several of the group 5 and 6 members are expressed up to, or within, the hindbrain. Many of these genes display varying levels of expression throughout the spinal column (not shown). Others, such as *Hoxa7* and *Hoxc8* display discrete domains, such that the posterior domain of expression is very weak relative to their anterior domains of expression. There are conflicting reports regarding the anterior borders of several *Hox* genes. For example, the anterior borders of *Hoxc9* and *Hoxc10* are reported at two different levels within the 12.5 dpc spinal column by two different groups (described in the text). These differences are indicated by the red outline. The anterior borders of *Hoxa11* and *Hoxa13* in the developing spinal column have not been reported at this stage.

approximately one rhombomere length posterior to that of *Hoxd4*.[81] This border does not appear to correspond to any rhombomere boundary, but starts near the beginning of the spinal column.

Within each group, there are differences in terms of timing and levels of expression. For example, *Hoxa1* expression proceeds *Hoxb1* expression, such that at 9.5 days post-coitum (dpc), the former is no longer expressed in the hindbrain.[16,77,78,80] Within the group 3 paralogs, *Hoxa3* and *Hoxb3* initially have identical patterns of expression but the anterior boundary of *Hoxb3* regresses posteriorly later in development.[82] They both display high levels of expression in r5 but have lower levels of expression in posterior domains. On the other hand, *Hoxd3* has weak expression in r5 but stronger expression in the posterior domains.[19] There are similar levels of variation within the group 4 paralogs. Although the anterior borders of *Hoxa4*, *Hoxb4* and *Hoxd4* are set at the r6/r7 boundary, their most anterior domain displays weaker expression, within r7, than their posterior domains of expression in the spinal column (Fig. 3).[13,19]

Between the groups, there are differences in how the anterior boundaries are set. The initial anterior border of *Hoxb2* expression persists at later stages and corresponds to the r2/r3 border. However, the degree to which the majority of *Hox* genes maintain or modulate their anterior borders is uncertain.[83,84] There is confusion because many have assumed that the expression domains represent a single pattern that is maintained or altered. However, regulatory analyses detailed below indicate that the overall patterns of *Hox* expression arise by the cumulative patterns mediated by a wide variety of individual local and global cis-regulatory elements and many of these individual sub-set of expression are highly variable.[27] Therefore, it is best to think of the patterns of *Hox* expression as continually dynamic rather than static in any way.

Hox *Genes from Groups 5 to 13 Begin Their Expression in the Developing Spinal Cord*

As with the *Hox* genes that begin their expression in the hindbrain, those that begin their expression in the spinal column display some general trends. The anterior border of most genes in groups 5 to 9 arise in the cervical region of the spinal cord while the anterior borders of groups 10 to 13 map to the lumbar region (Fig. 4).[85] There are slight differences reported between the clusters. *HoxA* genes 10 through 13 have slightly more anterior borders than paralogs 10 to 13 from the *HoxC* and *HoxD* clusters (Fig. 4).[85] *HoxB* members 5 through 9 have anterior borders that are more rostrally located in the cervical region of the spinal cord than their paralogs in the *HoxA*, *HoxC*, and *HoxD* clusters.[14,85] On the other hand, the anterior border of *Hoxb13* is more caudally located compared to group 13 paralogs from the other clusters.[86]

Group 1 to 4 *Hox* genes are also expressed in the spinal column, but their expression tends to be weaker than their expression in the hindbrain. For example, *Hoxb1*, *Hoxb2* and *Hoxb3* display higher levels of expression in the hindbrain but lower levels of expression in the spinal cord.[16] *Hoxb4* and *Hoxd4* have a similar level of expression throughout the spinal cord with slightly elevated levels in the dorsal part.[13] On the other hand, except for its most anterior domain, the expression of *Hoxa4* appears to be low throughout most of the spinal cord.[13]

Many early papers reported the domains of *Hox* gene expression at a uniform level throughout the spinal cord. However, more detailed studies have shown a variety of differences in both the D-V and A-P axes whereby discontinuous domains and varying levels occur in distinct regions of the spinal cord. For example, the expression pattern of *Hoxc4* gene displays a cyclic distribution pattern in the 12.5 dpc mouse CNS. It is strongly expressed from its anterior border near the end of the hindbrain down through the spinal column to an axial level that coincides with the adjacent developing 6th prevertebrae (pv). *Hoxc4* expression is relatively lower between pv6 and pv21, but again is detected at a higher level from pv21-pv29/pv30 before then decreasing in more posterior regions.[81] Another pattern had high levels in an anterior domain followed by a uniform lower level in more posterior regions. This is seen for both *Hoxa7* and *Hoxc8* expression (Fig. 4). The CNS domain of *Hoxa7* expression starts adjacent to pv3 and quickly fades at ~pv19 and is maintained at this lower level in the more posterior neural tube (Fig. 4).[87] Similarly, the spinal cord domain of *Hoxc8* expression starts at the level of pv4 and abruptly drops to low levels in the neural tube adjacent to pv14 (Fig. 4).[88] Hence, *Hox* genes can be expressed in a continuous domain that continues to the caudal end of the neural tube, in a restricted domain that begins and ends within the neural tube, and/or display areas of higher and lower expression along its length.

The anterior borders of *Hox* expression are not statically fixed in the spinal cord following their initial activation. This is particularly true for *Hox* members from groups 5 and 6 whose expression begins in the spinal column but spread forward at later stages to reside near or within the hindbrain by 12.5 dpc (Fig. 4). For example, from 9.5 to 11.5 dpc, the anterior borders of *Hoxb5*, *Hoxb6* and *Hoxb8* expand toward the hindbrain before reaching their anterior-most border.[89] At 11.5 dpc, *Hoxb5*'s anterior border resides within the posterior hindbrain while the anterior borders of *Hoxb6* and *Hoxb8* are level with the somite (s) boundaries s2/s3 and s4/s5, respectively.[89] Less dramatic changes have been reported for several group 9 to 13 members (*Abd-B* type). For example, the *Hoxc11* anterior border moves rostrally from the level of the second lumbar (L2) vertebrae at 12.5 dpc to reside adjacent to the level of L1 at 13.5 dpc.[90] So in addition to displaying dynamic levels of expression within A-P restricted domains, the initial anterior border of expression of many *Hox* genes may move more rostral as CNS development proceeds.

This dynamic repositioning of the anterior borders of expression in the developing neural tube helps to explain why there are discrepancies in the reporting of these borders. For example, the anterior border of *Hoxc9* in the 12.5 dpc mouse neural tube is reported to be at the level of pv10 or pv6.[88,91] Similarly, the anterior border of *Hoxc10* is either at the level of pv20 or pv18.[88,90] [b] Furthermore, the shifts in A-P boundaries mean that assigning precise borders at any given stage is also complicated by the need for accurate staging of embryos. Another interesting complication in determining A-P boundaries of expression is that many *Hox* genes produce alternate spliced products. Therefore, differences in the probes used for a given gene can detect different mRNA species with varying boundaries. This is observed for *Hoxa5* and *Hoxb3* in the spinal cord.[92,93]

While most studies on *Hox* expression have concentrated on RNA levels, the distribution of the encoded proteins is the actual domain of functional activity. It is interesting that post-translational regulation has shown differences in the posterior aspects of *Hox* expression as revealed by RNA in situ hybridization. In general anterior borders of expression as determined by RNA or antibody approaches are in agreement for any given gene. However, there often appears to be a sharp posterior border not seen with the probes for RNA. For example, the initial distribution of Hoxc8 protein in the spinal cord was examined at different stages by raising antibodies against the protein's unconserved amino-terminus.[94] In general, its protein's distribution occupied a smaller subset of the larger, relatively diffuse posterior distribution of its mRNA, being localized in the region that corresponds to the brachial domain of the spinal column. A more thorough examination of HoxC protein localization in the spinal cord was performed with antibodies against Hoxc5, Hoxc6, Hoxc8, Hoxc9 and Hoxc10.[95] Jessell and colleagues found that each Hox protein displayed discrete staggered and overlapping domains within the developing neural tube of the chick. In general, the protein distribution of the various HoxC proteins corresponds to the anterior end of their mRNA profile. Thus, the domain of Hoxc5 protein was the most anteriorly located, in the cervical region, while Hoxc10 protein was the most posteriorly located in the lumbar region, while Hoxc6, Hoxc8 and Hoxc9 proteins occupied distinct domains in between. An implication of these studies is that overlapping RNAs for *HoxC* genes does not necessarily correlate with overlapping protein expression, which has implications for understanding posterior prevalence.

The specific distribution of HoxC proteins is crucial for specifying "columns" of motor neurons produced at different regions along the spinal cord[96] in line with their dynamic D-V domains of restricted expression (see below). Thus Hoxc6 protein in the brachial region specifies LMC neurons that will go on to innervate the forelimb muscles, while Hoxc9 protein in the thoracic region specifies autonomic motor neurons that will innervate targets in the flank.[96] Misexpression of either gene in the other's domain of expression results in the down-regulation of the endogenous gene with a corresponding switch in MN columnar identity.[96] This is consistent with the idea that auto and cross-regulatory influences among the *Hox* genes are important for modulating their activity. Members from the other *Hox* clusters are also expressed in the spinal column, but only a handful of papers have looked at the effects that a single mutant,[97,98] compound mutants,[99-101] or misexpression[102,103] have on MN projections.

[b] These borders differences are highlighted in Figure 4 by the presence of a red rectangle.

Dorsal-Ventral (D-V) Patterning

The dynamic patterning of *Hox* gene expression in the A-P axis of the developing CNS is extended to the D-V axis as well. This is interesting as there is a precise ordered series of events associated with the birth of the major classes of neurons in the CNS.[104] As development proceeds over time from anterior to posterior the anterior regions of the spinal cord are more developmentally advanced than the posterior domains. In a comprehensive analysis of the D-V patterns of *HoxB* expression it was observed that all of the genes in the cluster are initially activated throughout the entire D-V domain, but as development proceeds they become dorsally-restricted in a manner that correlates with the formation of commissural and sensory neurons.[105] For example, at 10.5 dpc *Hoxb3* to *Hoxb9* are uniformly expressed throughout the D-V axis of the spinal cord.[105] But by 11.5 dpc, their expression intensifies in the lateral regions of the spinal cord where the relay neurons are being formed (Fig. 2C). The expression pattern of *Hoxb3, Hoxb6, Hoxb8* and *Hoxb9* then intensifies in the alar plates, with little or no expression of these genes in the basal plates. This strong dorsal expression forms an 'M' shaped-pattern, where the central domain of *HoxB* expression drops slightly ventrally into the ventricular layer past the level of the sulcus limitans. By 14.5 dpc, expression of *Hoxb8* and *Hoxb6* increases in the basal plates, particularly in the anterior portion of the neural tube, and matches the strong expression seen in the alar plates. But as one moves further posterior within the spinal cord, this ventral domain of expression fades, while the strong expression in the alar plates persists.

When Graham and coworkers (1991) compared the expression pattern of *Hoxb9* versus *Hoxc8* in 12.5 dpc mouse neural tube, they found that while *Hoxb9* exhibited the 'M' shaped pattern of strong dorsally restricted expression, *Hoxc8* had an entirely different pattern of expression stronger in ventral domains. It was excluded from a central 'V' shaped domain of expression within the transverse section of the neural tube, instead displaying a strong middle stripe of expression projecting laterally away from the border of the ventricular zone with lower levels of expression within the ventricular zone (Fig. 2D). This correlates with the birth of motor neurons and is in agreement with the more recent studies of Jessell and colleagues showing the important functional role of *HoxC* genes in specification of motorneuron fates.[33]

Although this was only a comparison between two genes from different clusters, it suggested that each complex might have a distinct D-V pattern of expression unique from the other *Hox* clusters, thereby forming a D-V combinatorial code for cell fates specification.[105,106] Further support for this idea come from additional expression studies. Unlike the strong, lateral stripe of *Hoxc8* expression, *Hoxc4*,[81] *Hoxc5* and *Hoxc6*[107] display robust expression throughout the basal plates of the neural tube, with a dorsal extension of this domain in the central region of the mantle layer into the alar plates. Except for some overlap in the D-V border, this pattern of expression is complementary to the pattern of expression observed for *HoxB* members in the spinal cord's D-V axis.[105] Knowing that the specific distribution of HoxC proteins along the A-P axis[95] specifies which 'column' of MNs will form from the ventral domain,[96] the distribution of *HoxB* protein in the dorsal domain may play an analogous role in determining the organization of the dorsal sensory area. Interestingly the targeted disruption of the *Hoxb8* gene results in mutant mice that display excessive, pathological grooming.[108] In trying to determine the mechanism of this behavior, Greer and Capecchi examined normal *Hoxb8* expression in the adult mouse and found that it is expressed in regions of the CNS known as the obsessive-compulsive disorder (OCD) circuit in humans. Loss of *Hoxb8* may have 'short-circuited' this neuronal network involved in innate behavior. It will be interesting to see what the effects that the loss of other *HoxB* members expressed within the spinal column has on behavioral-neuronal networks.

Regulation of Neural Expression

Given that their temporal and spatial expression must be precisely controlled, the regulation of *Hox* gene expression is understandably complex. It can be broken down into three stages: (1) initiation, (2) establishment, and (3) maintenance. Many cis-acting elements that control the initiation and establishment of *Hox* gene expression have been identified.[27,109] The majority of these involve the regulation of the members from the first four paralog group because of interest in understanding how expression is regulated and coupled to hindbrain segmentation. Regulatory analyses have revealed that there are multiple rhombomeric enhancers that act as independent regulatory elements. They direct expression in specific-subsets of hindbrain segments and function outside of the *Hox* complexes on heterologous promoters linked to reporter genes.[82,110-129] A variety of other regulatory elements that mediate neural expression in other domains have also been found.[89,92,112,116-118,129-137]

While these enhancers can display regulatory activity outside of the complex there is evidence that within a cluster some elements and *Hox* enhancers for other tissues may also be shared or competed for between adjacent genes.[93,130,136,138-140] An interesting but open question is the extent to which these enhancers may work more globally within a complex. In addition to these local transcriptional control elements, *Hox* expression can be controlled post-transcriptionally by the recently discovered *Hox* specific miRNAs.[141,142]

Local Regulators

The identification of local cis-regulatory enhancers capable of regulating segmental expression in the hindbrain opened the way for further analyses to identify upstream factors that serve to direct these patterns of express and help build a picture of the upstream regulatory network of *Hox* genes in hindbrain segmentation. Here, we will focus on progress made in identifying upstream regulators that initiate segmental patterns of *Hox* gene expression in the hindbrain and craniofacial development.

Activating Protein-2 (AP-2) Regulates Expression of *Hoxa2* in the Cranial Neural Crest

Ap-2 genes code for transcription factors that contain a proline-rich trans-activation domain, a dimerization domain, and a basic helix DNA-binding domain.[143] They bind to DNA as homodimers and heterodimers. Initially, *Ap-2* is expressed in extra-embryonic tissue,[144] and subsequently becomes expressed in premigratory neural crest cell, remaining actively expressed in these cells during their migration and differentiation.[144-146] Since *AP-2* and *Hox* genes are both expressed in cranial neural crest cells, it is interesting to find that the expression of *Hoxa2* in neural crest cells migrating into the second branchial arch is dependent upon an enhancer that contains and requires *Ap-2* binding sites.[120] *Ap-2* is expressed in crest cells that do not express *Hoxa2* indicating that mechanisms restricting its activity on the *Hoxa2* enhancer must be working in head development. To date *AP-2* binding sites have not been characterized for other *Hox* genes that are expressed in cranial neural crest.

kreisler (*kr*)

kreisler (*kr*) is a classic mouse mutation generated in an X-ray mutagenesis screen.[147] Affected mice have inner ear defects accompanied by posterior hindbrain malformations in the r5 and r6 territories.[148,149] The locus responsible for the *kr* phenotype was identified and it was found that the mutation was the result of a chromosomal inversion of the *kr* gene.[150] The *kr* gene encodes the KRML1 protein, a member of the Maf B/Zip protein family.[150] The expression of this gene normally starts at 8.0 dpc and rapidly decreases after 9.5 dpc.[150] It is strongly expressed in r5 and r6, with a sharp rostral border at r4/r5 boundary and a more diffuse caudal border in the vicinity of r6/r7. In addition, *kr* is expressed in the cranial neural crest derived from the caudal hindbrain but its expression is not detected in the otic vesicle that lies adjacent to r5 and r6. This pattern of expression correlates with the morphological defects seen in r5 and r6; however, patterning defects are also present in r3 and r4 which are outside the normal expression domains of *kr*.[115,151,152] A mutation in the zebrafish version of *kreisler* (*valentino*) also has been shown to have a conserved role in regulating hindbrain segmentation.[53,55,153]

In *kr* embryos, the caudal border of *Hoxb1* expression, normally at r4/r5 is fuzzy and extends more posteriorly. Similarly, the anterior borders of *Hoxb3* and *Hoxb4*, normally at r4/r5 and r6/r7 respectively, are poorly defined. Consequently, there are missing cranial nerves and the affected patterns of gene expression in the presumptive r5/r6 area suggests that these specific rhombomeres fail to form.[149,151] In the absence of this gene this territory adopts characteristics of r4 and r5 fails to form, r6 is present.[115,154]

A direct role for KRML1 in *Hox* gene expression was first shown for *Hoxb3* where enhancers in both mouse and chick responsible for r5-restricted expression of this gene were identified.[113] Subsequently, an r5/r6 enhancer for *Hoxa3* was also identified and found to be under direct regulatory control by KRML1.[115] Additional experimental evidence for a direct role of KRML1 in the regulation of group 3 paralogs in r5 and r6 was provided by the ectopic expression of *kr* in r3.[155] This induced expression of *Hoxa3* and *Hoxb3* in r3 thereby transforming it into an r5-like segment.[155]

Krox20

Krox20 encodes a zinc finger DNA-binding protein that is homologous to the *Drosophila* gap gene *Krüppel*. In mice, chickens, fish and *Xenopus* embryos, *Krox20* is transiently expressed in presumptive r3 and r5.[54,156,157] After the rhombomeres have formed, *Krox20* is sequentially down-regulated in r3 and then r5.[156,157] Targeted inactivation of *Krox20* causes mutant mice to die shortly after birth.[158,159] Histological examination reveals that nerves derived from the embryonic hindbrain are fused and disorganized. This defect is associated with a failure to maintain early r3 character and a failure to specify r5.[158-160]

Functional *Krox20* binding sites have been identified in enhancers of murine *Hoxa2*, *Hoxb2*, and *Hoxb3*.[114,124,125,128,161] The *Krox20*-dependent enhancers of *Hoxa2* and *Hoxb2* drive expression of these genes in r3 and r5[125,128] and are conserved in both chicken and pufferfish.[123,124] Regulation of *Hoxb3* expression in r5 requires both Krox20 and KRML1.[114] Krox20 binding sites have also been identified in the promoter regions of *Hoxa4*,[162] and human *HOXA7* and *HOXA9*.[163,164] However, none of these genes are expressed in r3 and r5. These Krox20 binding sites may regulate expression of these genes in other tissues where *Krox20* is expressed such as in the myelinating Schwann cells, chondrocytes, or osteoblasts. Krox20 has also been found to regulate the segmental expression of *EphA4* in r3 and r5 of the hindbrain.[165]

Nuclear Factor Y (NFY) and Ying-Yang 1 (YY1)

NFY is a ubiquitously expressed transcription factor[166,167] that interacts with the histone acetylase P/CAF[168] and the coactivators p300/CBP.[169] Hence, it acts as a transcriptional activator. On the other hand, YY1 is a zinc finger transcription factor that displays a dual nature, being capable of recruiting either histone acetyltransferases (HATs) or histone deacetylases (HDACs).[170] Since HATs acetylate amino-terminal histone tails thereby reducing the interactions between neighboring nucleosomes, they make DNA more accessible for transcription factors. On the other hand, HDACs deacetylate histone tails, making DNA more compact and less accessible. Together, these two factors, NFY and YY1, have been found to be important for a subdomain of *Hoxb4* expression in the neural tube.[171] More specifically, NFY binds to a region within *Hoxb4*'s Region C[117] and requires YY1 to stabilize its binding to the NFY binding site within this element.[171] Given the central role of HATs in gene activation and NFY's ubiquitous expression pattern, one would predict that other NFY binding sites exist in other *Hox* regulatory elements, but so far additional function sites have not been reported.

Sox and Oct

More than twenty *Sox* genes are found in vertebrates, the prototype of which is the *Sry* gene.[172] These genes code for proteins that bind to DNA through a high mobility group (HMG) domain. This domain interacts with the minor groove and induces a dramatic bend in the DNA. On its own, the binding of Sox to naturally occurring enhancers is often not enough to exert regulatory activity because many of these family members lack an activation domain.

Therefore to recruit coactivators there is frequently an adjacent DNA binding site. A common copartner for Sox appears to be Oct family members.[173-175] This Sox-Oct partnership appears to be relevant in the regulation of *Hoxb1* as there is an enhancer that mediates r4-restricted expression (the *Hoxb1* autoregulatory element, ARE)[110] and Sox and Oct cooperatively bind to a highly conserved bipartite binding site found within this element.[176] Mutation of this SOX-OCT binding site reduces reporter activity in P19 embryonal carcinoma (EC) cells but does not abolish r4-restricted expression of the transgene in mouse embryos.[176] Instead, the loss of this bipartite site alters the ability of the ARE to respond to retinoic acid.[176]

Global Regulators

Global regulators are factors or mechanistic features that simultaneously affect the expression of many *Hox* genes within a complex. In many instances, evidence for their direct action on *Hox* gene expression (i.e., the identification of cis-acting elements) has not been demonstrated. Two types of global regulators are major embryonic signaling pathways: the retinoic acid (RA) and fibroblast growth factor (FGF) signaling pathways. In a variety of tissues these signaling pathways act in an antagonistic fashion and it has been found that they differentially regulate *Hox* gene expression in the hindbrain and spinal cord. Their ability to regulate neural patterning can depend upon integration by a third global regulator, *Cdx*, a transcription factor whose expression is modulated by both of these signaling pathways.

Cdx Mutants Result in Posteriorization of Hox Anterior Expression Borders

Caudal (*cad*) was initially identified in *Drosophila*[177] where it is expressed in the fly embryo in a gradient that peaks at the posterior pole.[178] Using the *Drosophila cad* cDNA as a probe, a murine *caudal* ortholog, *Cdx1* was identified.[179] Subsequently, a total of three *Cdx* genes have been identified in vertebrates.[c] These genes code for proteins that have a highly conserved homeodomain and thus they function as transcription factors.

During early mouse embryogenesis, *Cdx2* and *Cdx4* are first expressed in extra-embryonic tissues,[180,181] whereas expression of *Cdx1* is first detected in the ectoderm and mesoderm of the primitive streak.[182] By 8.5 dpc, all three genes are expressed in an overlapping pattern in the neural tube and mesoderm. As the trunk of the embryo elongates and somitogenesis proceeds, their anterior borders of expression recede posteriorly. This dynamic pattern of expression in the trunk is conserved amongst the vertebrates [chick,[183] *Xenopus*,[184] zebrafish,[185,186]].

A role for *Cdx* in *Hox* gene regulation in the mouse was revealed by the targeted disruption of *Cdx1*.[187] These mutant mice display vertebral homeotic transformations that coincide with a posterior shift in the anterior borders of several *Hox* genes in the prevertebrae. In the homozygous state targeted disruption of *Cdx2* is embryonic lethal; however, heterozygous *Cdx2* mutants display homeotic transformations of the axial skeleton.[188] Products from both these genes were shown to have a synergistic role in setting *Hox* anterior borders when these two mouse mutants were crossed. Thus, compound mutants (*Cdx1*$^{-/-}$, *Cdx2*$^{+/-}$) have greater axial skeletal defects and greater posterior shifts in *Hox* gene expression than those present in single mutants.[189]

Work in *Xenopus* shows that over-expression of different *Xenopus cad* genes affects different subsets of *Hox* genes. For example, over-expression of *Xcad2* results in microencephaly and shortening of the A-P axis.[190] These morphological changes are accompanied by repression of *Hoxd1, Hoxb1, Hoxb3* and *Hoxb4* expression, and an anterior shift in the expression domains of *Hoxc6* and *Hoxb9*. On the other hand, over-expression of *Xcad3* does not affect *Hoxb1* and *Hoxb3* expression but does stimulate expression of *Hoxc6, Hoxa7, Hoxb7* and *Hoxb9*.[191] However, as with over-expression of *Xcad2*, it does anteriorize the borders of *Hoxa7* and *Hoxb9* resulting in embryos with anterior truncations.[191] When a dominant negative form of *Xcad3* is

[c] In the mouse, these are *Cdx1, Cdx2,* and *Cdx4*. The chicken and *Xenopus* orthologs are *CdxA, CdxB,* and *CdxC*; and *Xcad1, Xcad2,* and *Xcad3,* respectively.

over-expressed in *Xenopus* embryos, expression of *Hoxc6, Hoxa7, Hoxb7* and *Hoxb9* are blocked.[191] These studies show that there is a correlation between altered *Cdx* expression and *Hox* gene expression but do not establish a direct interaction.

In the mouse functional *Cdx* sites have been characterized in the mesodermal enhancers of *Hoxa5*,[192] *Hoxa7*,[193,194] *Hoxb8*,[195] and *Hoxc8* [196] thus arguing that Cdx directly affects *Hox* gene expression. A scan of the genomic sequences of many *Hox* genes reveals that they can contain one to four putative Cdx1 binding sites.[187] Although many of these putative Cdx sites may not be functional, the resulting shifts in the anterior borders of *Hox* genes in the *Cdx* mutants suggest that Cdx plays a central role in fine-tuning the establishment of these borders. The degree to which Cdx regulates neural expression is not clear. Most of the putative Cdx sites described above reside in mesodermal regulatory elements. However there is evidence in chick that Cdx may play a global role in regulating neural expression in the context of RA and FGF signaling.[197] We will further highlight the central role that Cdx plays as an integrator of FGF and RA signaling in section 5.3.

Retinoic Acid (RA) Signaling

Early experiments in which pregnant rats were maintained on vitamin A-deficient (VAD) diets resulted in fetuses with malformations in their visual, respiratory, vascular and urogenital systems.[198-200] These defects could be reversed by supplementing the VAD diet with vitamin A. However, the timing of its administration is important to prevent a given malformation.[200] At the time, these results suggested that vitamin A is important for early developmental events. We now know that retinoic acid (RA) is the active derivative of vitamin A.

Too much or too little RA results in embryological defects. When excess RA is applied during gastrulation and neurulation, embryos display abnormal brain, head and vertebral development.[24,201-209] These abnormalities include anterior displacement of the otic vesicles, expansion of the anterior hindbrain and spinal cord, and the formation of the somites at more anterior locations. Conversely, reducing or eliminating RA signaling results in the partial transformation of the rhombomeres or the complete loss of the posterior hindbrain and anterior spinal cord depending on the severity of the perturbation to the RA signaling pathway.[210-215] In either case, the resulting abnormal patterning of the hindbrain in these embryos coincided with misexpression of the 3' *Hox* genes in this structure.[202,213,216,217]

Early cell culture experiments showed that *Hox* genes display another form of colinearity associated with RA signaling: genes at the 3' end of the complex respond rapidly to RA while genes positioned in more 5' positions respond slowly.[218-220] For several of these *Hox* genes, their response to RA is directly mediated through retinoic acid response elements (RAREs) identified in their adjacent sequences. Functional RAREs have been identified for group 1,[111,112,221-223] group 4,[116,129,131,224,225] and recently for group 5 *Hox* genes.[226] In transgenic mice assays the individual mutation of several of these RAREs reduces or eliminates neural expression of the reporter,[112,116,224,225] arguing that they play a pivotal role in the endogenous gene's RA-mediated CNS expression. Unexpectedly, the genetic deletion or mutation of these neural-specific RAREs only delays and diminishes their cognate *Hox* gene's expression producing mice that are viable and fertile.[70,227] Even the generation of the compound mutant of these altered RAREs produces only mild facial motor nerve defects and results in low perinatal lethality amongst the homozygous compound mutants.[67] One possible explanation for the discrepancy between the RARE transgenic reporter results and their mutation within the endogenous locus is that other cis-acting elements such as auto- and cross-regulatory enhancers compensate for the loss of these individual RAREs. RA excess or depletion can produce the gross rearrangements of the hindbrain with altered *Hox* gene expression profiles because these compensatory mechanisms are over-ridden when all *Hox* genes are misregulated.

Although initial in vitro work suggested that all *Hox* genes are up-regulated by RA, in the embryo 5' members may be actually negatively regulated by RA signaling.[95] In early chick neural tube explants, adding RA to the explants did not induce *Hoxc6, Hoxc8, Hoxc9* nor

Hoxc10 expression in the developing MNs or adjacent cells. However, when these explants are exposed to FGF-signaling and subsequent RA-signaling, *Hoxc6* expression is elevated by 2 folds whereas *Hoxc8* and *Hoxc9* expression are decreased by 3 and 6 folds, respectively. The authors concluded that RA signaling (from the paraxial mesoderm) refines the pattern of *HoxC* expression induced in the MNs by FGF signaling.[95] Similar differences in the response of 3' and 5' *HoxB* genes to RA in the CNS have been reported.[197] Hence, RA signaling regulates the anterior border in a positive fashion in the hindbrain, whereas in the spinal column it appears to function in an opposite manner, limiting the anterior expansion of more posteriorly expressed *Hox* genes. The presence of RAREs clearly indicates that many of the influences of RA may be direct on the *Hox* genes via their RAR and RXR nuclear receptors.

Fibroblast Growth Factor (FGF) Signaling

For more than a decade, it has been known that exogenously added FGF can up-regulate expression of posteriorly expressed *Hox* genes in the mesoderm and neural tissues of *Xenopus* embryos.[228-234] Conversely, blocking FGF signaling in *Xenopus* embryos, by expressing a dominant-negative FGF receptor, results in repression or posteriorization of *Hox* gene expression.[234,235] These observations suggest that endogenous FGF signaling controls A-P patterning in part through regulation of *Hox* gene expression. A growing body of work now shows that this is a common mechanism during vertebrate development.

The fibroblast growth factor (FGFs) family contains twenty-two secreted heparin-binding molecules that interact with one of four FGF receptors (FGFRs).[236] This signaling pathway is further complicated by the observations that three of FGFRs have various isoforms, and that these isoforms display different preferences for FGF ligands.[237] Most of the *Fgf* genes have been mutated in mouse gene targeting experiments and result in a variety of phenotypes.[238] Of these, the targeted deletion of *Fgf2, Fgf3, Fgf8*, and *Fgf14* have been reported to affect some aspect of CNS development.

In the early zebrafish hindbrain, both *Fgf3* and *Fgf8* are transiently expressed. Their expression precedes the expression of ***kreisler*** and *Krox20* and becomes refined to the region that will form r4.[239,240] When the rhombomeres form, the expression of *Fgf3* persists in r4 while *Fgf8* expression in this region disappears. Expression of *Fgf8* is subsequently detected at the midbrain/hindbrain border.[241,242] The patterns of *Fgf* expression vary among vertebrates suggesting that different members of the family may have distinct roles in each species. Inhibition of FGF signaling in the early chick and fish hindbrain reduces and blocks expression of ***kreisler*** and *Krox20*.[240,243] In the zebrafish, this inhibition leads to the failure of *Hoxa2* expression while the initial expression of *Hoxb1* becomes down-regulated at later stages. These changes result in a zebrafish hindbrain in which the axonal organization is severely affected.[240] Conversely, exogenous applied FGF, or its ectopic expression, in the vertebrate hindbrain induces ectopic expression of ***kreisler*** and *Krox20*.[239,243] The expression of *Fgf* in the presumptive r4 of zebrafish has been used to suggest that it may act as a signaling center for hindbrain patterning by regulating the upstream effectors of *Hox* gene expression but this pattern is not conserved amongst all vertebrates.[244-246]

In the early hindbrain FGF signaling positively regulates *Hox* gene expression, but at later stages, its role is reversed, as *Fgf8* expression from the isthmus represses *Hox* gene expression thereby limiting the anterior border of *Hoxa2* expression at the r1/r2 border.[247] However, at the opposite end of the embryo, FGF signaling from the regressing node activates group 5 to group 13 *Hox* genes in the spinal cord.[197] This apparent paradox is the result of Cdx availability which is no longer present in the hindbrain when these genes are active.[197]

In chick neural tube explants, increasing concentrations of FGF induce expression of more 5' located *HoxC* genes.[95] At low concentrations of Fgf8 or Fgf2, chick neural plate explants express *Hoxc6*. Higher concentrations of Fgf8 or Fgf2 induce *Hoxc8* and *Hoxc9* (in addition to *Hoxc6*), and at the highest concentration they also induce *Hoxc10* expression. Conversely, blocking FGF signaling with SU5402, an inhibitor of FGFR1 activity, prevent the expression of

Hoxc8, Hoxc9 and *Hoxc10* in chick neural explants. Furthermore, the presence of SU5402 is able to block the *HoxC* inducing activities of Hensen node that were cocultured with the chick explants. Higher concentrations of SU5402 are needed to inhibit expression of *Hoxc6*. Thus the expression of *Hoxc6* to *Hoxc10* in MNs relies on FGF signaling provided by Hensen's node. Similarly it has been shown that FGF signaling is required for expression of 5' *HoxB* genes in the neural tube.[197] These data suggest that FGF signaling acts in a reciprocal fashion to RA signaling by promoting expression of *Hox* genes in the spinal column.

Tying It All Together: RA and FGF Signaling Regulate Hox *Gene Expression in the CNS*

So far, we have reviewed how RA signaling can initiate expression of the 3' *Hox* genes (from groups 1 to 5) while FGF signaling can induce expression of the 5' *Hox* genes (from groups 6 to 13) in the developing vertebrate CNS. At the boundary between these two groups, group 6 *Hox* members can be induced by both signaling pathways [95,197]. Recent work has highlighted how these two signaling pathways play a pivotal role in promoting neuron formation in the developing spinal column and how they act in a reciprocal and antagonistic process to do so.[248,249] RA signaling from the somitic mesoderm represses *Fgf8* expression in the developing spinal cord thus restricting its expression to the caudal end of the embryo. This repressive interaction is necessary for neuronal differentiation to proceed within the neural tube adjacent to the forming somites.[248] Paradoxically FGF signaling initiates *Raldh2* expression in the paraxial mesoderm, resulting in the somites that flank the neural tube to produce RA.[249] In both *Xenopus* and chick, *caudal* genes are downstream targets of FGF signaling[191,197,234] and in the chick neural tube *Cdx* is required to anteriorize the FGF response of 5' *HoxB* members into the hindbrain.[197] Moreover, RA signaling can directly regulate the expression of *Cdx* genes.[250] Hence if FGFs and RA themselves global regulators can modulate another global regulator, *Cdx*, there is a complex network that regulates the balance of these opposing pathways. This might help explain their roles in neurogenesis and the differential responsiveness of the 3' *Hox* genes and 5' *Hox* genes to RA and FGF signaling. There is also evidence that *Cdx* genes may integrate signals from the Wnt pathway, which regulates A-P patterning.[251-255] By simultaneously linking the establishment of A-P character in both the paraxial mesoderm (from which the muscles are generated) and neural tube, there is a coordinated system whereby the muscles and neurons that innervate them have the same molecular code.

Expression and Neuronal Phenotypes

The majority of *Hox* loss-of-function and gain-of-function studies describe the effects that these mutations have on axial and limb skeleton patterning and formation. In terms of what effects these mutations have on CNS development, more work has been produced looking at the defects in the organization of the hindbrain and its cranial nerves; whereas, a smaller body of work deals with the alterations and deletions of the MNs projecting from the spinal column into the limbs or trunk.

In the hindbrain targeted inactivation of the *Hoxa1* gene results in loss of r5 and reduction in the size of r4.[65,256] However, the loss of Hoxb1 function is mild in comparison. Initially r4 character is triggered but it is not maintained and there is a transformation to an r2-like character.[69,257] The generation of compound group 1 mutants leads to more severe phenotypes. Hence in *Hoxa1* and *Hoxb1* double homozygous mutants r4 and r5 fail to be specified and are missing along with derivatives of the second branchial arch which are generated from neural crest cells that migrate from r4.[61,67,70] As a consequence of their gain or loss of expression in the hindbrain, *Hoxb1* and *Hoxa1* affect the projections of the cranial MNs from the hindbrain.[59,60,66,69,258] These studies have shown that Hoxb1 is essential for the formation, migration and projections of the facial MN.[59,69,257,259] Studies with the *Hoxa1, Hoxb1* and *Hoxb2* mutants also show later effects of these genes on neurogenesis.[68,260] Similar work with group 3 paralogs have highlighted their important role in the formation of the abducens and hypoglossal neurons.[63,261]

Forcing expression of *Hox* genes in more anterior domains of the hindbrain typically leads to rhombomere transformations. Ectopic expression of *Hoxa1* results in transformation of r2 into r4,[56,60] and similar results are produced when *Hoxa1* and *Hoxb1* expression is anteriorized in the hindbrain by excess RA treatment.[206,217] These changes in rhombomere identity have profound effects on the subsequently formed CNS structures, such as the cerebellum that is derived primarily from r1.[262] Loss of *Hoxa2* causes a caudal expansion of the cerebellum,[66] which is further extended into r2 and r3 territory when both paralogous *Hoxa2* and *Hoxb2* are absent.[260]

The products of *Hox* genes also determine the axonal projection patterns from the spinal column. Targeted inactivation of *HoxA, HoxC* and *HoxD* genes results in alterations and/or reductions of the MNs axonal projections to muscle targets in the limbs.[97-100,263] In general, members from groups 6 and 8 are important for specifying brachial MNs that innervate the forelimbs whereas members from groups 9, 10 and 11 have been shown to control MN projections into the hindlimbs. Interestingly, mutation of the *Hoxb13* gene does not produce rearrangements or deletions of elements derived from the CNS but results in the overgrowth of the posterior spinal cord and tail vertebrae.[264] This suggests that, unlike the other *Abd-B* like Hox proteins, Hoxb13 functions as a repressor of neuronal and caudal vertebral proliferation.

Studies in which the gene is mis-expressed in the target area, or in tissue through which the neurons project, have highlighted the importance of coordinating the simultaneous expression of these genes in both the paraxial mesoderm and neuroectoderm. Viral misexpression of *Hoxc6* in the chick cervical paraxial mesoderm disrupts the spinal nerve's projection, causing it to prematurely halt is migration.[102] This suggests that the signals from the mesoderm through which the axon migrates are important to direct their outgrowth.[102] Similarly, the viral misexpression of *Hoxb1* in the first branchial arches causes *Hoxb1*-specified neurons to alter their migration patterns so that they project axons into the first branchial arch rather than their normal targets located in the second branchial arch.[59] This link between the local A-P character of the neural tube and the flanking mesoderm tissue is missed in gene-targeting experiments because such experiments eliminated the targeted gene's product from both compartments.

The Roles of Auto- and Cross-Regulatory Interactions between *Hox* Genes

An important consideration in both the regulation and function of *Hox* genes in the nervous system is the degree to which they cross-regulate each other. A common feature from the cis-regulatory studies on segmental expression of *Hox* genes in the hindbrain is that early signals (RA, FGFs) and transcription factors (kreisler, Krox20) are transient. Hence, other mechanisms are required to maintain or stabilize their domains of expression critical for later expression and function in rhombomeric segments. By analogy to *Drosophila*, mechanisms involving Polycomb and trithorax mediated epigenetic changes in chromatin, have been widely postulated to be important in vertebrate *Hox* regulation. In support of this, mouse mutants for members of the *PcG* and *trxG* have been shown to alter *Hox* expression, primarily examined in the mesoderm due to the prevalence of skeletal defects.[265] However, the regulatory analyses of rhombomeric enhancers has uncovered a surprising degree of auto and cross-regulation between the *Hox* genes themselves, as an important mechanism for maintaining segmentally-restricted expression.[70,82,110,116,122,123,140,266-268] During the establishment and maintenance stages, Hox proteins can directly cross-regulate the expression of other *Hox* genes,[82,122,140,269-271] or perpetuate their own expression through auto-regulatory elements.[82,110,267,271] Hence, following initiation by transiently expressed upstream factors or signals, Hox response elements serve to generate feedback loops that stabilize and perpetuate segmental expression. These lock-down or feedback loops have important implications in functional studies because mutation of one *Hox* gene can alter the regulation of other family members, generating a more complex phenotype.

One of the best examples of such loops from regulatory and mutant analyses relates to rhombomere 4. It appears that there is a pathway for specifying r4 character and later neuronal identities dependent upon cross-talk between *Hox* genes.[268] Early retinoid signaling triggers both *Hoxb1* and *Hoxa1* expression directly via RAREs located at the 3' end of these genes.[56,112,217,227] In combination with Pbx and Prep/Meis cofactors, both Hoxa1 and Hoxb1 then bind to the auto-regulatory enhancer of *Hoxb1* to maintain its expression.[110,272-278] Hoxb1 in turn directly cross regulates both *Hoxb2* and *Hoxa2* specific expression in r4 and Hoxb2 feed-backs upon *Hoxb1* to support its expression.[110,112,122,123,268] By these loops it is clear that *Hoxb1* plays a central role in specifying and maintaining r4 identity and the mutation of any of the individual *Hox* genes in this these cross-talk loops directly correlates with the neuronal phenotypes observed.[268]

Further evidence for the importance of these feedback loops in other vertebrates is provided by loss and gain-of-function experiments in zebrafish where *Hox*, *Pbx* and *Meis* mutations all show primary defects in hindbrain patterning and alterations of *Hox* expression.[51,55,58,279-283] There is also emerging evidence that both positive and negative feedback loops among *Hox* genes is important in the spinal cord.[33,96]

In conclusion, regulatory analyses seeking to understand the cis-regulation of *Hox* genes themselves and the upstream cascade have surprisingly provided novel insight into potential *Hox* target genes. By identifying a series of known in vivo relevant target elements it will be possible to define the cis-regulatory code of a Hox response element and use it to predict down-stream target genes identified by genomic and microarray analysis. Therefore, a major area of interest in future *Hox* research will be to understand how different Hox proteins bind their target sites, what cofactors they use and how the output of activation or repression is achieved. This will be important in predicting and understanding their interactions on down-stream target genes. This work should also be highly relevant to *Hox* gene expression, regulation and function in tissues outside the nervous system.

References

1. Lewis EB. A gene complex controlling segmentation in Drosophila. Nature 1978; 276:565-570.
2. McGinnis W, Garber RL, Wirz J et al. A homologous protein-coding sequence in Drosophila homeotic genes and its conservation in other metazoans. Cell 1984; 37:403-408.
3. Carrasco AE, McGinnis W, Gehring WJ et al. Cloning of an X. laevis gene expressed during early embryogenesis coding for a peptide region homologous to Drosophila homeotic genes. Cell 1984; 37:409-414.
4. Scott M, Weiner A, Hazelrigg T et al. The molecular organization of the Antennapedia locus of Drosophila. Cell 1983; 35:763-776.
5. McGinnis W, Krumlauf R. Homeobox genes and axial patterning. Cell 1992; 68:283-302.
6. Krumlauf R. Hox genes in vertebrate development. Cell 1994; 78:191-201.
7. Graham A, Papalopulu N, Hunt P et al. The murine and Drosophila homeobox clusters are derived from a common ancestor based on similarities in structure and expression. In: Capecchi M, ed. Molecular Genetics of Early Drosophila and Mouse Development. Cold Spring Harbor: Cold Spring Harpor Press, 1989:85-90.
8. Duboule D, Dollé P. The structural and functional organization of the murine HOX gene family resembles that of Drosophila homeotic genes. EMBO J 1989; 8(5):1497-1505.
9. Amores A, Force A, Yan YL et al. Zebrafish hox clusters and vertebrate genome evolution. Science 1998; 282:1711-1714.
10. Amores A, Suzuki T, Yan YL et al. Developmental roles of pufferfish Hox clusters and genome evolution in ray-fin fish. Genome Res 2004; 14(1):1-10.
11. Hoegg S, Meyer A. Hox clusters as models for vertebrate genome evolution. Trends Genet 2005; 21:421-424.
12. Aparicio S, Hawker K, Cottage A et al. Organization of the Fugu rubripes Hox clusters, evidence for continuing evolution of vertebrate Hox complexes. Nat Gen 1997; 16:79-84.
13. Gaunt SJ, Krumlauf R, Duboule D. Mouse homeo-genes within a subfamily, Hox-1.4, -2.6 and -5.1, display similar anteroposterior domains of expression in the embryo, but show stage- and tissue-dependent differences in their regulation. Development 1989; 107(1):131-141.
14. Graham A, Papalopulu N, Krumlauf R. The murine and Drosophila homeobox gene complexes have common features of organization and expression. Cell 1989; 57:367-378.

15. Dollé P, Izpisùa-Belmonte JC, Falkenstein H et al. Coordinate expression of the murine Hox-5 complex homeobox-containing genes during limb pattern formation. Nature 1989; 342(6251):767-772.
16. Wilkinson DG, Bhatt S, Cook M et al. Segmental expression of Hox-2 homeobox-containing genes in the developing mouse hindbrain. Nature 1989; 341:405-409.
17. Izpisua-Belmonte JC, Falkenstein H, Dollé P et al. Murine genes related to the Drosophila AbdB homeotic genes are sequentially expressed during development of the posterior part of the body. EMBO J 1991; 10:2279-2289.
18. Hunt P, Wilkinson D, Krumlauf R. Patterning the vertebrate head: Murine Hox 2 genes mark distinct subpopulations of premigratory and migrating neural crest. Development 1991; 112:43-51.
19. Hunt P, Gulisano M, Cook M et al. A distinct Hox code for the branchial region of the vertebrate head. Nature 1991; 353(6347):861-864.
20. Dollé P, Izpisua-Belmonte JC, Falkenstein H et al. Coordinate expression of the murine Hox-5 complex homoeobox-containing genes during limb pattern formation. Nature 1989; 342:767-772.
21. Izpisua-Belmonte J, Falkenstein H, Dollé P et al. Murine genes related to the Drosophila AbdB homeotic gene are sequentially expressed during development of the posterior part of the body. EMBO J 1991; 10:2279-2289.
22. Nohno T, Noji S, Koyama E et al. Involvement of the Chox-4 chicken homeobox genes in determination of anteroposterior axial polarity during limb development. Cell 1991; 64:1197-1205.
23. Dollé P, Izpisùa-Belmonte JC, Brown JM et al. Hox-4 genes and the morphogenesis of mammalian genitalia. Genes Dev 1991; 5:1767-1776.
24. Kessel M, Gruss P. Homeotic transformations of murine prevertebrae and concommitant alteration of Hox codes induced by retinoic acid. Cell 1991; 67:89-104.
25. Kessel M, Balling R, Gruss P. Variations of cervical vertebrae after expression of a Hox 1.1 transgene in mice. Cell 1990; 61:301-308.
26. Gehring WJ, Qian YQ, Billeter M et al. Homeodomain-DNA recognition. Cell 1994; 78:211-223.
27. Maconochie M, Nonchev S, Morrison A et al. Paralogous Hox genes: Function and regulation. Ann Rev Genet 1996; 30:529-556.
28. Favier B, Dolle P. Developmental functions of mammalian Hox genes. Mol Hum Reprod 1997; 3:115-131.
29. Slack JMW, Tannahill D. Mechanism of anteroposterior axis specification in vertebrates. Lessons from the amphibians. Development 1992; 114:285-302.
30. Nieuwkoop P. Activation and organization of the central nervous system in amphibians. J Exp Zool 1952; 120:1-108.
31. Nieuwkoop P. Inductive interactions in early amphibian development and their general nature. J Embryol Exp Morph 1985; 89(suppl.):333-347.
32. Lumsden A, Krumlauf R. Patterning the vertebrate neuraxis. Science 1996; 274:1109-1115.
33. Jessell TM. Neuronal specification in the spinal cord: Inductive signals and transcriptional codes. Nat Rev Genet 2000; 1(1):20-29.
34. Le Douarin N, Kalcheim C. The Neural Crest. 2nd ed. Cambridge Univesity Press, 1999.
35. Keynes R, Krumlauf R. Hox genes and regionalization of the nervous system. Ann Rev Neurosci 1994; 17:109-132.
36. Lumsden A, Keynes R. Segmental patterns of neuronal development in the chick hindbrain. Nature 1989; 337:424-428.
37. Fraser S, Keynes R, Lumsden A. Segmentation in the chick embryo hindbrain is defined by cell lineage restrictions. Nature 1990; 344:431-435.
38. Akam M, Dawson I, Tear G. Homeotic genes and the control of segment diversity. Development 1988; 104(Supplement: Mechanisms of Segmentation):123-133.
39. Chisaka O, Capecchi MR. Regionally restricted developmental defects resulting from targeted disruption of the mouse homeobox gene hox1.5. Nature 1991; 350:473-479.
40. Condie BG, Capecchi MR. Mice homozygous for a targeted disruption of Hoxd-3(Hox-4.1) exhibit anterior transformations of the first and second cervical vertebrae, the atlas and axis. Development 1993; 119:579-595.
41. Gendron-Maguire M, Mallo M, Zhang M et al. Hoxa-2 mutant mice exhibit homeotic transformation of skeletal elements derived from cranial neural crest. Cell 1993; 75:1317-1331.
42. Jeannottee L, Lemieux M, Charron J et al. Specification of axial identity in the mouse: Role of the Hoxa-5(Hox1.3) gene. Genes Dev 1993; 7:2085-2096.
43. Rijli FM, Mark M, Lakkaraju S et al. A homeotic transformation is generated in the rostral branchial region of the head by disruption of Hoxa-2, which acts as a selector gene. Cell 1993; 75:1333-1349.
44. Ramirez-Solis R, Zheng H, Whiting J et al. Hoxb-4 (Hox-2.6) mutant mice show homeotic transformation of a cervical vertebra and defects in the closure of the sternal rudiments. Cell 1993; 73:279-294.

45. Small KS, Potter S. Homeotic transformations and limb defects in Hoxa-11 mutant mice. Genes Dev 1993; 7:2318-2328.
46. Jegalian BG, De Robertis EM. Homeotic transformations in the mouse induced by over-expression of a human Hox3.3 transgene. Cell 1992; 71:901-910.
47. Le Mouellic H, Lallemand Y, Brulet P. Homeosis in the mouse induced by a null mutation in the Hox-3.1 gene. Cell 1992; 69:251-264.
48. Birgbauer E, Fraser SE. Violation of cell lineage restriction compartments in the chick hindbrain. Development 1994; 120:1347-1356.
49. Clarke JD, Lumsden A. Segmental repetition of neuronal phenotype sets in the chick embryo hindbrain. Development 1993; 118:151-162.
50. Clarke JDW, Erskine L, Lumsden A. Differential progenitor dispersal and the spatial origin of early neurons can explain the predominance of single-phenotype clones in the chick hindbrain. Dev Dyn 1998; 212:14-26.
51. McClintock JM, Kheirbek MA, Prince VE. Knockdown of duplicated zebrafish hoxb1 genes reveals distinct roles in hindbrain patterning and a novel mechanism of duplicate gene retention. Development 2002; 129(10):2339-2354.
52. McClintock JM, Carlson R, Mann DM et al. Consequences of Hox gene duplication in the vertebrates: An investigation of the zebrafish Hox paralogue group 1 genes. Development 2001; 128(13):2471-2484.
53. Moens CB, Cordes SP, Giorgianni MW et al. Equivalence in the genetic control of hindbrain segmentation in fish and mouse. Development 1998; 125:381-391.
54. Moens CB, Kimmel CB. Hindbrain patterning in the zebrafish embryo. Soc Neurosci Abstr 1995; 21:118.118.
55. Moens CB, Prince VE. Constructing the hindbrain: Insights from the zebrafish. Dev Dyn 2002; 224(1):1-17.
56. Zhang M, Kim HJ, Marshall H et al. Ectopic Hoxa-1 induces rhombomere transformation in mouse hindbrain. Development 1994; 120:2431-2442.
57. Prince VE, Joly L, Ekker M et al. Zebrafish hox genes: Genomic organization and modified colinear expression patterns in the trunk. Development 1998; 125:407-420.
58. Prince VE, Moens CB, Kimmel CB et al. Zebrafish hox genes: Expression in the hindbrain region of wild-type and mutants of the segmentation gene, valentino. Development 1998; 125:393-406.
59. Bell E, Wingate RJ, Lumsden A. Homeotic transformation of rhombomere identity after localized Hoxb1 misexpression. Science 1999; 284:21682171.
60. Alexandre D, Clarke J, Oxtoby E et al. Ectopic expression of Hoxa-1 in the zebrafish alters the fate of the mandibular arch neural crest and phenocopies a retinoic acid -induced phenotype. Development 1996; 122:735-746.
61. Rossel M, Capecchi MR. Mice mutant for both Hoxa1 and Hoxb1 show extensive remodeling of the hindbrain and defects in craniofacial development. Development 1999; 126:5027-5040.
62. Gaufo GO, Wu S, Capecchi MR. Contribution of Hox genes to the diversity of the hindbrain sensory system. Development 2004; 131(6):1259-1266.
63. Gaufo GO, Thomas KR, Capecchi MR. Hox3 genes coordinate mechanisms of genetic suppression and activation in the generation of branchial and somatic motoneurons. Development 2003; 130:5191-5201.
64. Lufkin T, Dierich A, LeMeur M et al. Disruption of the Hox-1.6 homeobox gene results in defects in a region corresponding to its rostral domain of expression. Cell 1991; 66(6):1105-1119.
65. Mark M, Lufkin T, Vonesch JL et al. Two rhombomeres are altered in Hoxa-1 mutant mice. Development 1993; 119:319-338.
66. Gavalas A, Davenne M, Lumsden A et al. Role of Hoxa-2 in axon pathfinding and rostral hindbrain patterning. Development 1997; 124:3693-3702.
67. Gavalas A, Studer M, Lumsden A et al. Hoxa1 and Hoxb1 synergize in patterning the hindbrain, cranial nerves and second pharyngeal arch. Development 1998; 125:1123-1136.
68. Gavalas A, Ruhrberg C, Livet J et al. Neuronal defects in the hindbrain of Hoxa1, Hoxb1 and Hoxb2 mutants reflect regulatory interactions among these Hox genes. Development 2003.
69. Studer M, Lumsden A, Ariza-McNaughton L et al. Altered segmental identity and abnormal migration of motor neurons in mice lacking Hoxb-1. Nature 1996; 384:630-635.
70. Studer M, Gavalas A, Marshall H et al. Genetic interaction between Hoxa1 and Hoxb1 reveal new roles in regulation of early hindbrain patterning. Development 1998; 125:1025-1036.
71. Lim T, Jaques K, Stern C et al. An evaluation of myelomeres and segmentation of the chick embryo spinal cord. Development 1991; 113:227-238.
72. Stern C, Jaques K, Lim T et al. Segmental lineage restrictions in the chick embryo spinal cord depend on the adjacent somites. Development 1991; 113:239-244.

73. Phelan KA, Hollyday M. Axon guidance in muscleless chick wings: The role of muscle cells in motoneuronal pathway selection and muscle nerve formation. J Neurosci 1990; 10:2699-2716.
74. Ensini M, Tsuchida T, Belting HG et al. The control of rostrocaudal pattern in the developing spinal cord: Specification of motor neuron subtype identity is initiated by signals from paraxial mesoderm. Development 1998; 125:969-982.
75. Bronner-Fraser M. Rostrocaudal differences within the somites confer segmental pattern to trunk neural crest migration. Curr Top Dev Biol 2000; 47:279-296.
76. Krumlauf R, Holland P, McVey J et al. Developmental and spatial patterns of expression of the mouse homeobox gene, Hox 2.1. Development 1987; 99:603-617.
77. Murphy P, Hill RE. Expression of the mouse labial-like homeobox-containing genes, Hox 2.9 and Hox 1.6, during segmentation of the hindbrain. Development 1991; 111:61-74.
78. Murphy P, Davidson DR, Hill RE. Segment-specific expression of a homeobox-containing gene in the mouse hindbrain. Nature 1989; 341:156-159.
79. Prince V, Lumsden A. Hox-a2 expression in normal and transposed rhombomeres: Independent regulation in the neural tube and neural crest. Development 1994; 120:911-923.
80. Frohman MA, Boyle M, Martin GR. Isolation of the mouse Hox-2.9 gene; analysis of embryonic expression suggests that positional information along the anterior-posterior axis is specified by mesoderm. Development 1990; 110:589-607.
81. Geada AMC, Gaunt SJ, Azzawi M et al. Sequence and embryonic expression of the murine Hox-3.5 gene. Development 1992; 116:497-506.
82. Manzanares M, Bel-Vialer S, Ariza-McNaughton L et al. Independent regulation of initiation and maintenance phases of Hoxa3 expression in the vertebrate hindbrain involves auto and cross-regulatory mechanisms. Development 2001; 128(18):3595-3607.
83. Forlani S, Lawson KA, Deschamps J. Acquisition of Hox codes during gastrulation and axial elongation in the mouse embryo. Development 2003; 130:3807-3819.
84. Deschamps J, van Nes J. Developmental regulation of the Hox genes during axial morphogenesis in the mouse. Development 2005; 132:2931-2942.
85. Carpenter EM. Hox genes and spinal cord development. Dev Neurosci 2002; 24:24-34.
86. Zeltser L, Desplan C, Heintz N. Hoxb-13: Colinear expression of a new Hox gene in a distant region of the HOXB cluster. Development 1996; 122:2475-2484.
87. Mahon KA, Westphal H, Gruss P. Expression of homeobox gene Hox 1.1 during mouse embryogenesis. Development 1988; 104(Suppl):187-195.
88. Peterson RL, Jacobs DF, Awgulewitsch A. Hox-3.6: Isolation and characterization of a new murine homeobox gene located in the 5' region of the Hox-3 cluster. Mech Dev 1992; 37:151-166.
89. Oosterveen T, Meijlink F, Deschamps J. Expression of retinaldehyde dehydrogenase II and sequential activation of 5' Hoxb genes in the mouse caudal hindbrain. Gene Expr Patterns 2004; 4:243-247.
90. Hostikka SL, Capecchi MR. The mouse Hoxc11 gene: Genomic structure and expression pattern. Mech Dev 1998; 70:133-145.
91. Erselius JR, Goulding MD, Gruss P. Structure and expression pattern of the murine Hox-3.2 gene. Development 1990; 110:629-642.
92. Larochelle C, Tremblay M, Bernier D et al. Multiple cis-acting regulatory regions are required for restricted spatio-temporal Hoxa5 gene expression. Dev Dyn 1999; 214:127-140.
93. Sham MH, Hunt P, Nonchev S et al. Analysis of the murine Hox-2.7 gene: Conserved alternative transcripts with differential distributions in the nervous system and the potential for shared regulatory regions. EMBO J 1992; 11:1825-1836.
94. Awgulewitsch A, Jacobs D. Differential expression of Hox 3.1 protein in subregions of the embryonic and adult spinal cord. Development 1990; 108(3):411-420.
95. Liu JP, Laufer E, Jessell TM. Assigning the positional identity of spinal motor neurons. rostrocaudal patterning of Hox-c expression by FGFs, Gdf11, and Retinoids. Neuron 2001; 32(6):997-1012.
96. Dasen JS, Liu JP, Jessell TM. Motor neuron columnar fate imposed by sequential phases of Hox-c activity. Nature 2003; 425(6961):926-933.
97. Carpenter EM, Goddard JM, Davis AP et al. Targeted disruption of Hoxd-10 affects mouse hindlimb development. Development 1997; 124:4505-4514.
98. Tiret L, Le Mouellic H, Maury M et al. Increased apoptosis of motoneurons and altered somatotopic maps in the branchial spinal cord of Hoxc8-deficient mice. Development 1998; 125:279-291.
99. de la Cruz CC, Der-Avakian A, Spyropoulos DD et al. Targeted disruption of Hoxd9 and Hoxd10 alters locomotor behavior, vertebral identity, and peripheral nervous system development. Dev Biol 1999; 216:595-610.
100. Lin AW, Carpenter EM. Hoxa10 and Hoxd10 coordinately regulate lumbar motor neuron patterning. J Neurobiol 2003; 56:328-337.

101. Wahba GM, Hostikka SL, Carpenter EM. The paralogous Hox genes Hoxa10 and Hoxd10 interact to pattern the mouse hindlimb peripheral nervous system and skeleton. Dev Biol 2001; 231:87-102.
102. Burke AC, Tabin CJ. Virally mediated misexpression of Hoxc-6 in the cervical mesoderm results in spinal nerve truncations. Dev Biol 1996; 178:192-197.
103. Shah V, Drill E, Lance-Jones C. Ectopic expression of Hoxd10 in thoracic spinal segments induces motoneurons with a lumbosacral molecular profile and axon projections to the limb. Dev Dyn 2004; 231:43-56.
104. Altman J, Bayer J. The development of the rat spinal cord. Adv Anat Embryol Cell Biol 1984; 85:1-166.
105. Graham A, Maden M, Krumlauf R. The murine Hox-2 genes display dynamic dorsoventral patterns of expression during central nervous system development. Development 1991; 112:255-264.
106. Gaunt SJ. Expression patterns of mouse Hox genes: Clues to an understanding of developmental and evolutionary strategies. Bioessays 1991; 13(10):505-513.
107. Gaunt SJ, Coletta PL, Pravtcheva D et al. Mouse Hox-3.4: Homeobox sequence and embryonic expression patterns compared with other members of the Hox gene network. Development 1990; 109(2):329-339.
108. Greer JM, Capecchi MR. Hoxb8 is required for normal grooming behavior in mice. Neuron 2002; 33:23-34.
109. Lufkin T. Transcriptional regulation of vertebrate Hox genes during embryogenesis. Crit Rev Eukaryot Gene Expr 1997; 7:195-213.
110. Pöpperl H, Bienz M, Studer M et al. Segmental expression of Hoxb1 is controlled by a highly conserved autoregulatory loop dependent upon exd/Pbx. Cell 1995; 81:1031-1042.
111. Studer M, Pöpperl H, Marshall H et al. Role of a conserved retinoic acid response element in rhombomere restriction of Hoxb-1. Science 1994; 265:1728-1732.
112. Marshall H, Studer M, Pöpperl H et al. A conserved retinoic acid response element required for early expression of the homeobox gene Hoxb-1. Nature 1994; 370:567-571.
113. Manzanares M, Cordes S, Kwan CT et al. Segmental regulation of Hoxb3 by kreisler. Nature 1997; 387:191-195.
114. Manzanares M, Nardelli J, Gilardi-Hebenstreit P et al. Krox20 and kreisler cooperate in the transcriptional control of segmental expression of Hoxb3 in the developing hindbrain. EMBO J 2002; 21(3):365-376.
115. Manzanares M, Cordes S, Ariza-McNaughton L et al. Conserved and distinct roles of kreisler in regulation of the paralogous Hoxa3 and Hoxb3 genes. Development 1999; 126:759-769.
116. Gould A, Itasaki N, Krumlauf R. Initiation of rhombomeric Hoxb4 expression requires induction by somites and a retinoid pathway. Neuron 1998; 21:39-51.
117. Whiting J, Marshall H, Cook M et al. Multiple spatially specific enhancers are required to reconstruct the pattern of Hox-2.6 gene expression. Genes Dev 1991; 5(11):2048-2059.
118. Morrison A, Ariza-McNaughton L, Gould A et al. HOXD4 and regulation of the group 4 paralog genes. Development 1997; 124:3135-3146.
119. Frasch M, Chen X, Lufkin T. Evolutionary-conserved enhancers direct region-specific expression of the murine Hoxa-1 and Hoxa-2 loci in both mice and Drosophila. Development 1995; 121:957-974.
120. Maconochie M, Krishnamurthy R, Nonchev S et al. Regulation of Hoxa2 in cranial neural crest cells involves members of the AP-2 family. Development 1999; 126:1483-1494.
121. Maconochie MK, Nonchev S, Manzanares M et al. Differences in Krox20-dependent regulation of Hoxa2 and Hoxb2 during hindbrain development. Dev Biol 2001; 233(2):468-481.
122. Maconochie MK, Nonchev S, Studer M et al. Cross-regulation in the mouse HoxB complex: The expression of Hoxb2 in rhombomere 4 is regulated by Hoxb1. Genes Dev 1997; 11:1885-1896.
123. Tumpel S, Cambronero F, Wiedemann LM et al. Evolution of cis elements in the differential expression of two Hoxa2 coparalogous genes in pufferfish (Takifugu rubripes). Proc Natl Acad Sci USA 2006; 103:5419-5424.
124. Nonchev S, Maconochie M, Vesque C et al. The conserved role of Krox-20 in directing Hox gene expression during vertebrate hindbrain segmentation. Proc Natl Acad Sci USA 1996; 93:9339-9345.
125. Nonchev S, Vesque C, Maconochie M et al. Segmental expression of Hoxa-2 in the hindbrain is directly regulated by Krox-20. Development 1996; 122:543-554.
126. Zhang F, Pöpperl H, Morrison A et al. Elements both 5' and 3' to the murine Hoxd4 gene establish anterior borders of expression in mesoderm and neuroectoderm. Mech Dev 1997; 67:49-58.
127. Vesque C, Maconochie M, Nonchev S et al. Hoxb-2 transcriptional activation in rhombomeres 3 and 5 requires an evolutionarily conserved cis-acting element in addition to the Krox-20 binding site. EMBO J 1996; 15:5383-5896.

128. Sham MH, Vesque C, Nonchev S et al. The zinc finger gene Krox-20 regulates Hoxb-2 (Hox2.8) during hindbrain segmentation. Cell 1993; 72:183-196.
129. Packer AI, Crotty DA, Elwell VA et al. Expression of the murine Hoxa4 gene requires both autoregulation and a conserved retinoic acid response element. Development 1998; 125:1991-1998.
130. Sharpe J, Nonchev S, Gould A et al. Selectivity, sharing and competitive interactions in the regulation of Hoxb genes. EMBO J 1998; 17:1788-1798.
131. Pöpperl H, Featherstone M. Identification of a retinoic acid repsonse element upstream of the murine Hox-4.2 gene. Mol Cell Biol 1993; 13(1):257-265.
132. Morrison A, Chaudhuri C, Ariza-McNaughton L et al. Comparative analysis of chicken Hoxb-4 regulation in transgenic mice. Mech Dev 1995; 53:47-59.
133. Packer AI, Mailutha KG, Ambrozewicz LA et al. Regulation of the Hoxa4 and Hoxa5 genes in the embryonic mouse lung by retinoic acid and TGFbeta1: Implications for lung development and patterning. Dev Dyn 2000; 217(1):62-74.
134. Behringer R, Crotty DA, Tennyson VM et al. Seguences 5' of the homeobox of the Hox-1.4 gene direct tissue-specific expression of lacZ during mouse development. Development 1993; 117:823-833.
135. Nowling T, Zhou W, Krieger KE et al. Hoxa5 gene regulation: A gradient of binding activity to a brachial spinal cord element. Dev Biol 1999; 208:134-146.
136. Moreau J, Jeannotte L. Sequence analysis of a Hoxa4-Hoxa5 intergenic region including shared regulatory elements. DNA Seq 2002; 13:203-209.
137. Tuggle CK, Zakany J, Cianetti L et al. Region-specific enhancers near two mammalian homeo box genes define adjacent rostrocaudal domains in the central nervous system. Genes Dev 1990; 4(2):180-189.
138. Zakany J, Gérard M, Favier B et al. Deletion of a HoxD enhancer induces transcriptional heterochrony leading to transposition of the sacrum. EMBO J 1997; 16:4393-4402.
139. Gérard M, Chen JW, Gronemeyer H et al. In vivo targeted mutagenesis of a regulatory element reguired for positioning the Hoxd-11 and Hoxd-10 expression boundaries. Genes Dev 1996; 10:2326-2334.
140. Gould A, Morrison A, Sproat G et al. Positive cross-regulation and enhancer sharing: Two mechanisms for specifying overlapping Hox expression patterns. Genes Dev 1997; 11:900-913.
141. Mansfield JH, Harfe BD, Nissen R et al. MicroRNA-responsive 'sensor' transgenes uncover Hox-like and other developmentally regulated patterns of vertebrate microRNA expression. Nat Genet 2004.
142. Hornstein E, Mansfield JH, Yekta S et al. The microRNA miR-196 acts upstream of Hoxb8 and Shh in limb development. Nature 2005; 438:671-674.
143. Williams T, Tjian R. Analysis of the DNA-binding and activation properties of the human transcription factor AP-2. Genes Dev 1991; 5:670-682.
144. Mitchell PJ, Timmons PM, Hébert JM et al. Transcription factor AP-2 is expressed in neural crest cell lineages during mouse embryogenesis. Genes Dev 1991; 5:105-119.
145. Moser M, Ruschoff J, Buettner R. Comparative analysis of AP-2α and AP-2β gene expression during mouse embryogenesis. Dev Dyn 1997; 208:115-124.
146. Chazaud C, Oulad-Abdelghani M, Bouillet P et al. AP-2.2, a novel gene related to AP-2, is expressed in the forebrain, limbs and face during mouse embryogenesis. Mech Dev 1996; 54:83-94.
147. Deol MS. The abnormalities of the inner ear in kreisler mice. J Embryol Exp Morphol 1964; 12:475-490.
148. McKay I, Lewis J, Lumsden A. Organization and development of facial motor neurons in the kreisler mutant mouse. Eur J Neurosci 1997; 9:1499-1506.
149. McKay IJ, Muchamore I, Krumlauf R et al. The kreisler mouse: A hindbrain segmentation mutant that lacks two rhombomeres. Development 1994; 120:2199-2211.
150. Cordes SP, Barsh GS. The mouse segmentation gene kr encodes a novel basic domain-leucine zipper transcription factor. Cell 1994; 79:1025-1034.
151. Frohman MA, Martin GR, Cordes SP et al. Altered rhombomere-specific gene expression and hyoid bone differentiation in the mouse segmentation mutant, kreisler (kr). Development 1993; 117:925-936.
152. Sadl VS, Sing A, Mar L et al. Analysis of hindbrain patterning defects caused by the kreisler(enu) mutation reveals multiple roles of Kreisler in hindbrain segmentation. Dev Dyn 2003; 227:134-142.
153. Moens CB, Yan YL, Appel B et al. Valentino: A zebrafish gene required for normal hindbrain segmentation. Development 1996; 122:3981-3990.
154. Giudicelli F, Gilardi-Hebenstreit P, Mechta-Grigoriou F et al. Novel activities of Mafb underlie its dual role in hindbrain segmentation and regional specification. Dev Biol 2003; 253(1):150-162.
155. Theil T, Ariza-McNaughton L, Manzanares M et al. Requirement for downregulation of kreisler during late patterning of the hindbrain. Development 2002; 129(6):1477-1485.

156. Nieto MA, Bradley LC, Wilkinson DG. Conserved segmental expression of Krox-20 in the vertebrate hindbrain and its relationship to lineage restriction. Development Supplement 1991; 2:59-62.
157. Wilkinson DG, Bhatt S, Chavrier P et al. Segment-specific expression of a zinc finger gene in the developing nervous system of the mouse. Nature 1989; 337:461-464.
158. Schneider-Maunoury S, Topilko P, Seitandou T et al. Disruption of Krox-20 results in alteration of rhombomeres 3 and 5 in the developing hindbrain. Cell 1993; 75:1199-1214.
159. Swiatek PJ, Gridley T. Perinatal lethality and defects in hindbrain development in mice homozygous for a targeted mutation of the zinc finger gene Krox-20. Genes Dev 1993; 7:2071-2084.
160. Voiculescu O, Taillebourg E, Pujades C et al. Hindbrain patterning: Krox20 couples segmentation and specification of regional identity. Development 2001; 128(24):4967-4978.
161. Vogel A, Tickle C. FGF-4 maintains polarizing activity of posterior limb buds cells in vivo and in vitro. Development 1993; 119(1):199-206.
162. Chavrier P, Vesque C, Galliot B et al. The segment-specific gene Krox-20 encodes a transcription factor with binding sites in the promoter of the Hox 1.4 gene. EMBO J 1990; 9:1209-1218.
163. Kim MH, Chang HH, Shin C et al. Genomic structure and sequence analysis of human HOXA-9. DNA Cell Biol 1998; 17:407-414.
164. Kim MH, Cho M, Park D. Sequence analysis of the 5'-flanking region of the gene encoding human HOXA-7. Somat Cell Mol Genet 1998; 24:371-374.
165. Theil T, Frain M, Gilardi-Hebenstreit P et al. Segmental expression of the EphA4 (Sek-1) receptor tyrosine kinase in the hindbrain is under direct transcriptional control of Krox-20. Development 1998; 125(3):443-452.
166. McNabb DS, Xing Y, Guarente L. Cloning of yeast HAP5: A novel subunit of a heterotrimeric complex required for CCAAT binding. Genes Dev 1995; 9:47-58.
167. Sinha S, Maity SN, Lu J et al. Recombinant rat CBF-C, the third subunit of CBF/NFY, allows formation of a protein-DNA complex with CBF-A and CBF-B and with yeast HAP2 and HAP3. Proc Natl Acad Sci USA 1995; 92:1624-1628.
168. Bertagna A, Jahroudi N. The NFY transcription factor mediates induction of the von Willebrand factor promoter by irradiation. Thromb Haemost 2001; 85:837-844.
169. Faniello MC, Bevilacqua MA, Condorelli G et al. The B subunit of the CAAT-binding factor NFY binds the central segment of the Coactivator p300. J Biol Chem 1999; 274:7623-7626.
170. Thomas MJ, Seto E. Unlocking the mechanisms of transcription factor YY1: Are chromatin modifying enzymes the key? Gene 1999; 236:197-208.
171. Gilthorpe J, Vandromme M, Brend T et al. Spatially specific expression of Hoxb4 is dependent on the ubiquitous transcription factor NFY. Development 2002; 129:3887-3899.
172. Kamachi Y, Uchikawa M, Kondoh H. Pairing SOX off: With partners in the regulation of embryonic development. Trends Genet 2000; 16:182-187.
173. Botquin V, Hess H, Fuhrmann G et al. New POU dimer configuration mediates antagonistic control of an osteopontin preimplantation enhancer by Oct-4 and Sox-2. Genes Dev 1998; 12(13):2073-2090.
174. Ambrosetti DC, Scholer HR, Dailey L et al. Modulation of the activity of multiple transcriptional activation domains by the DNA binding domains mediates the synergistic action of Sox2 and Oct-3 on the fibroblast growth factor-4 enhancer. J Biol Chem 2000; 275(30):23387-23397.
175. Boyer LA, Lee TI, Cole MF et al. Core transcriptional regulatory circuitry in human embryonic stem cells. Cell 2005; 122:947-956.
176. Di Rocco G, Gavalas A, Popperl H et al. The recruitment of SOX/OCT complexes and the differential activity of HOXA1 and HOXB1 modulate the Hoxb1 auto-regulatory enhancer function. J Biol Chem 2001; 1:1.
177. Mlodzik M, Fjose A, Gehring WJ. Isolation of caudal, a Drosophila homeo box-containing gene with maternal expression, whose transcripts form a concentration gradient at the preblastoderm stage. EMBO J 1985; 4:2961-2969.
178. Mlodzik M, Gehring WJ. Expression of the caudal gene in the germ line of Drosophila: Formation of an RNA and protein gradient during early embryogenesis. Cell 1987; 48:465-478.
179. Duprey P, Chowdhury K, Dressler GR et al. A mouse gene homologous to the Drosophila gene caudal is expressed in epithelial cells from the embryonic intestine. Genes Dev 1988; 2:1647-1654.
180. Gamer LW, Wright CV. Murine Cdx-4 bears striking similarities to the Drosophila caudal gene in its homeodomain sequence and early expression pattern. Mech Dev 1993; 43:71-81.
181. Beck F, Erler T, Russell A et al. Expression of Cdx-2 in the mouse embryo and placenta: Possible role in patterning of the extra-embryonic membranes. Dev Dyn 1995; 204:219-227.
182. Meyer BI, Gruss P. Mouse Cdx-1 expression during gastrulation. Development 1993; 117:191-203.
183. Marom K, Shapira E, Fainsod A. The chicken caudal genes establish an anterior-posterior gradient by partially overlapping temporal and spatial patterns of expression. Mech Dev 1997; 64:41-52.

184. Pillemer G, Epstein M, Blumberg B et al. Nested expression and sequential downregulation of the Xenopus caudal genes along the anterior-posterior axis. Mech Dev 1998; 71:193-196.
185. Davidson AJ, Ernst P, Wang Y et al. cdx4 mutants fail to specify blood progenitors and can be rescued by multiple hox genes. Nature 2003; 425:300-306.
186. Joly JS, Maury M, Joly C et al. Expression of a zebrafish caudal homeobox gene correlates with the establishment of posterior cell lineages at gastrulation. Differentiation 1992; 50:75-87.
187. Subramanian V, Meyer BI, Gruss P. Disruption of the murine homeobox gene Cdx1 affects axial skeletal identities by altering the mesodermal expression domains of Hox genes. Cell 1995; 83(4):641-653.
188. Chawengsaksophak K, James R, Hammond VE et al. Homeosis and intestinal tumours in Cdx2 mutant mice. Nature 1997; 386:84-87.
189. van den Akker E, Forlani S, Chawengsaksophak K et al. Cdx1 and Cdx2 have overlapping functions in anteroposterior patterning and posterior axis elongation. Development 2002; 129:2181-2193.
190. Epstein M, Pillemer G, Yelin R et al. Patterning of the embryo along the anterior-posterior axis: The role of the caudal genes. Development 1997; 124:3805-3814.
191. Isaacs H, Pownall M, Slack J. Regulation of Hox gene expression and posterior development by the Xenopus caudal homolog Xcad3. EMBO J 1998; 17:3413-3427.
192. Tabaries S, Lapointe J, Besch T et al. Cdx protein interaction with Hoxa5 regulatory sequences contributes to Hoxa5 regional expression along the axial skeleton. Mol Cell Biol 2005; 25:1389-1401.
193. Knittel T, Kessel M, Kim MH et al. A conserved enhancer of the human and murine Hoxa-7 gene specifies the anterior boundary of expression during embryonal development. Development 1995; 121:1077-1088.
194. Min W, Cho M, Jang SI et al. Sequence and functional analysis of an upstream regulatory region of human HOXA7 gene. Gene 1996; 182:1-6.
195. Charité J, de Graaff W, Consten D et al. Transducing positional information to the Hox henes: Critical interaction of cdx gene products with position-sensitive regulatory elements. Development 1998; 125:4349-4358.
196. Shashikant CS, Ruddle FH. Combinations of closely situated cis-acting elements determine tissue-specific patterns and anterior extent of early Hoxc8 expression. Proc Natl Acad Sci USA 1996; 93:12364-12369.
197. Bel-Vialar S, Itasaki N, Krumlauf R. Initiating Hox gene expression: In the early chick neural tube differential sensitivity to FGF and RA signaling subdivides the HoxB genes in two distinct groups. Development 2002; 129(22):5103-5115.
198. Wilson JG, Warkany J. Congenital anomalies of heart and great vessels in offspring of vitamin A-deficient rats. Am J Dis Child 1950; 79:963.
199. Wilson JG, Warkany J. Malformations in the genito-urinary tract induced by maternal vitamin A deficiency in the rat. Am J Anat 1948; 83:357-407.
200. Wilson JG, Roth CB, Warkany J. An analysis of the syndrome of malformations induced by maternal vitamin A deficiency. Effects of restoration of vitamin A at various times during gestation. Am J Anat 1953; 92:189-217.
201. Morriss GM. Morphogenesis of the malformations induced in rat embryos by maternal hypervitaminosis A. J Anat 1972; 113:241-250.
202. Morriss-Kay GM, Murphy P, Hill RE et al. Effects of retinoic acid excess on expression of Hox-2.9 and Krox-20 and on morphological segmentation in the hindbrain of mouse embryos. EMBO J 1991; 10:2985-2995.
203. Simeone A, Avantaggiato V, Moroni MC et al. Retinoic acid induces stage-specific antero-posterior transformation of rostral central nervous system. Mech Dev 1995; 51(1):83-98.
204. Papalopulu N, Clarke J, Bradley L et al. Retinoic acid causes abnormal development and segmental patterning of the anterior hindbrain in Xenopus embryos. Development 1991; 113:1145-1159.
205. Leonard L, Horton C, Maden M et al. Anteriorization of CRABP-I expression by retinoic acid in the developing mouse central nervous system and its relationship to teratogenesis. Dev Biol 1995; 168:514-528.
206. Hill J, Clarke JDW, Vargesson N et al. Exogenous retinoic acid causes specific alterations in the development of the midbrain and hindbrain of the zebrafish embryo including positional respecification of the Mauthner neuron. Mech Dev 1995; 50:3-16.
207. Holder N, Hill J. Effects of retinoic acid on zebrafish embryos: Modulation of engrailed protein expression at the midbrain-hindbrain border and development of cranial ganglia. Development 1991; 113:1159-1170.
208. Durston AJ, Timmermans JP, Hage WJ et al. Retinoic acid causes an anteroposterior transformation in the developing central nervous system. Nature 1989; 340:140-144.
209. Sive H, Draper B, Harland R et al. Identification of a retinoic acid-sensitive period during primary axis formation in Xenopus laevis. Genes Dev 1990; 4:932-942.

210. Niederreither K, Vermot J, Schuhbaur B et al. Retinoic acid synthesis and hindbrain patterning in the mouse embryo. Development 2000; 127(1):75-85.
211. White JC, Highland M, Kaiser M et al. Vitamin A-deficiency in the rat embryo results in anteriorization of the posterior hindbrain which is prevented by maternal consumption of retinoic acid or retinol. Dev Biol 2000; 220:263-284.
212. Maden M, Gale E, Kostetskii I et al. Vitamin A deficient quail embryos have half a hindbrain and other neural defects. Curr Biol 1996; 6:417-426.
213. Dupe V, Lumsden A. Hindbrain patterning involves graded responses to retinoic acid signalling. Development 2001; 128(12):2199-2208.
214. Dupé V, Ghyselinck NB, Wendling O et al. Key roles of retinoic acid receptors alpha and beta in the patterning of the caudal hindbrain, pharyngeal arches and otocyst in the mouse. Development 1999; 126:5051-5059.
215. Blumberg B, Bolado J, Moreno T et al. An essential role for retinoid signaling in anteroposterior neural patterning. Development 1997; 124:373-379.
216. Conlon RA, Rossant J. Exogenous retinoic acid rapidly induces anterior ectopic expression of murine Hox-2 genes in vivo. Development 1992; 116(2):357-368.
217. Marshall H, Nonchev S, Sham MH et al. Retinoic acid alters hindbrain Hox code and induces transformation of rhombomeres 2/3 into a 4/5 identity. Nature 1992; 360:737-741.
218. Simeone A, Acampora D, Arcioni L et al. Sequential activation of HOX2 homeobox genes by retinoic acid in human embryonal carcinoma cells. Nature 1990; 346:763-766.
219. Papalopulu N, Lovell-Badge R, Krumlauf R. The expression of murine Hox-2 genes is dependent on the differentiation pathway and displays a collinear sensitivity to retinoic acid in F9 cells and Xenopus embryos. Nuc Acids Res 1991; 19(20):5497-5506.
220. Simeone A, Acampora D, Nigro V et al. Differential regulation by retinoic acid of the homeobox genes of the four HOX loci in human embryonal carcinoma cells. Mech Dev 1991; 33:215-227.
221. Langston AW, Gudas LJ. Identification of a retinoic acid responsive enhancer 3' of the murine homeobox gene Hox-1.6. Mech Dev 1992; 38(3):217-228.
222. Langston A, Thompson J, Gudas L. Retinoic acid-responsive enhancers located 3' of the HoxA and the HoxB gene clusters. J Biol Chem 1997; 272:2167-2175.
223. Ogura T, Evans R. Evidence for two distinct retinoic acid response pathways for Hoxb-1 gene regulation. Proc Natl Acad Sci USA 1995; 92:392-396.
224. Morrison A, Moroni M, Ariza-McNaughton L et al. In vitro and transgenic analysis of a human HOXD4 retinoid-responsive enhancer. Development 1996; 122:1895-1907.
225. Zhang F, Nagy Kovacs E, Featherstone MS. Murine hoxd4 expression in the CNS requires multiple elements including a retinoic acid response element. Mech Dev 2000; 96(1):79-89.
226. Oosterveen T, Niederreither K, Dolle P et al. Retinoids regulate the anterior expression boundaries of 5' Hoxb genes in posterior hindbrain. EMBO J 2003; 22:262-269.
227. Dupé V, Davenne M, Brocard J et al. In vivo functional analysis of the Hoxa1 3' retinoid response element (3' RARE). Development 1997; 124:399-410.
228. Ruiz-i-Altaba A, Melton DA. Interaction between peptide growth factors and homeobox genes in the establishment of anteroposterior polarity in frog embryos. Nature 1989; 341:33-38.
229. Cho K, De Robertis E. Differential activation of Xenopus homeobox genes by mesoderm-inducing growth factors and retinoic acid. Genes Dev 1990; 4:1910-1917.
230. Kengaku M, Okamoto H. bFGF as a possible morphogen for the anteroposterior axis of the central nervous system in Xenopus. Development 1995; 121:3121-3130.
231. Cox WG, Hemmati-Brivanlou A. Caudalization of neural fate by tissue recombination and bFGF. Development 1995; 121:4349-4358.
232. Kolm P, Sive H. Regulation of the Xenopus labial homeodomain genes, HoxA1 and HoxD1: Activation by retinoids and peptide growth factors. Dev Biol 1995; 167:34-49.
233. Lamb TM, Harland RM. Fibroblast growth factor is a direct neural inducer, which combined with noggin generates anterior-posterior neural pattern. Development 1995; 121:3627-3636.
234. Pownall M, Tucker A, Slack J et al. eFGF, Xcad3 and Hox genes form a molecular pathway that establishes the anteroposterior axis in Xenopus. Development 1996; 122:3881-3892.
235. Godsave SF, Durston AJ. Neural induction and patterning in embryos deficient in FGF signaling. Int J Dev Biol 1997; 41(1):57-65.
236. Ornitz DM, Itoh N. Fibroblast growth factors. Genome Biol 2001; 2, (REVIEWS3005).
237. Ornitz DM, Xu J, Colvin JS et al. Receptor specificity of the fibroblast growth factor family. J Biol Chem 1996; 271:15292-15297.
238. Itoh N, Ornitz DM. Evolution of the Fgf and Fgfr gene families. Trends Genet 2004; 20:563-569.
239. Maves L, Jackman W, Kimmel CB. FGF3 and FGF8 mediate a rhombomere 4 signaling activity in the zebrafish hindbrain. Development 2002; 129(16):3825-3837.

240. Walshe J, Maroon H, McGonnell IM et al. Establishment of hindbrain segmental identity requires signaling by FGF3 and FGF8. Curr Biol 2002; 12(13):1117-1123.
241. Heikinheimo M, Lawshe A, Shackleford G et al. Fgf-8 expression in the post-gastrulation mouse suggests roles in the development of face, limbs and central nervous system. Mech Dev 1994; 48:129-138.
242. Crossley PH, Martin GR. The mouse FGF-8 gene encodes a family of polypeptides expressed in regions that direct outgrowth and patterning in the developing embryo. Development 1995; 121:439-451.
243. Marin F, Charnay P. Hindbrain patterning: FGFs regulate Krox20 and mafB/kr expression in the otic/preotic region. Development 2000; 127(22):4925-4935.
244. Mahmood R, Kiefer P, Guthrie S et al. Multiple roles for FGF-3 during cranial neural development in the chicken. Development 1995; 121(5):1399-1410.
245. Mahmood R, Mason IJ, Morriss-Kay GM. Expression of Fgf-3 in relation to hindbrain segmentation, otic pit position and pharyngeal arch morphology in normal and retinoic acid-exposed mouse embryos. Anat Embryol (Berl) 1996; 194:13-22.
246. Lombardo A, Isaacs HV, Slack JM. Expression and functions of FGF-3 in Xenopus development. Int J Dev Biol 1998; 42:1101-1107.
247. Irving C, Mason I. Signalling by FGF8 from the isthmus patterns anterior hindbrain and establishes the anterior limit of Hox gene expression. Development 2000; 127:177-186.
248. Diez del Corral R, Breitkreuz DN, Storey KG. Onset of neuronal differentiation is regulated by paraxial mesoderm and requires attenuation of FGF signalling. Development 2002; 129:1681-1691.
249. Diez del Corral R, Olivera-Martinez I, Goriely A et al. Opposing FGF and retinoid pathways control ventral neural pattern, neuronal differentiation, and segmentation during body axis extension. Neuron 2003; 40(1):65-79.
250. Houle M, Prinos P, Iulianella A et al. Retinoic acid regulation of Cdx1: An indirect mechanism for retinoids and vertebral specification. Mol Cell Biol 2000; 20(17):6579-6586.
251. Lohnes D. The Cdx1 homeodomain protein: An integrator of posterior signaling in the mouse. Bioessays 2003; 25(10):971-980.
252. Gaunt SJ, Drage D, Trubshaw RC. cdx4/lacZ and cdx2/lacZ protein gradients formed by decay during gastrulation in the mouse. Int J Dev Biol 2005; 49:901-908.
253. Gaunt SJ, Cockley A, Drage D. Additional enhancer copies, with intact cdx binding sites, anteriorize Hoxa-7/lacZ expression in mouse embryos: Evidence in keeping with an instructional cdx gradient. Int J Dev Biol 2004; 48:613-622.
254. Gaunt SJ, Drage D, Cockley A. Vertebrate caudal gene expression gradients investigated by use of chick cdx-A/lacZ and mouse cdx-1/lacZ reporters in transgenic mouse embryos: Evidence for an intron enhancer. Mech Dev 2003; 120:573-586.
255. Gaunt SJ. Gradients and forward spreading of vertebrate Hox gene expression detected by using a Hox/lacZ transgene. Dev Dyn 2001; 221:26-36.
256. Carpenter EM, Goddard JM, Chisaka O et al. Loss of Hoxa-1 (Hox-1.6) function results in the reorganization of the murine hindbrain. Development 1993; 118:1063-1075.
257. Goddard J, Rossel M, Manley N et al. Mice with targeted disruption of Hoxb1 fail to form the motor nucleus of the VIIth nerve. Development 1996; 122:3217-3228.
258. Jungbluth S, Bell E, Lumsden A. Specification of distinct motor neuron identities by the singular activities of individual Hox genes. Development 1999; 126(12):2751-2758.
259. Arenkiel BR, Tvrdik P, Gaufo GO et al. Hoxb1 functions in both motoneurons and in tissues of the periphery to establish and maintain the proper neuronal circuitry. Genes Dev 2004; 18(13):1539-1552.
260. Davenne M, Maconochie MK, Neun R et al. Hoxa2 and Hoxb2 control dorsoventral patterns of neuronal development in the rostral hindbrain. Neuron 1999; 22:677-691.
261. Guidato S, Prin F, Guthrie S. Somatic motoneurone specification in the hindbrain: The influence of somite-derived signals, retinoic acid and Hoxa3. Development 2003; 130:2981-2996.
262. Wingate RJ, Hatten ME. The role of the rhombic lip in avian cerebellum development. Development 1999; 126:4395-4404.
263. Wahba GM, Hostikka SL, Carpenter EM. The paralogous Hox genes Hoxa10 and Hoxd10 interact to pattern the mouse hindlimb peripheral nervous system and skeleton. Dev Biol 2001; 231(1):87-102.
264. Economides KD, Zeltser L, Capecchi MR. Hoxb13 mutations cause overgrowth of caudal spinal cord and tail vertebrae. Dev Biol 2003; 256:317-330.
265. Gould A. Functions of mammalian Polycomb-group and trithorax-group related genes. Curr Opin Gen Dev 1997; 7:488-494.

266. Tümpel S, Maconochie M, Wiedemann LM et al. Conservation and diversity in the cis-regulatory networks that integrate information controlling expression of Hoxa2 in hindbrain and cranial neural crest cells in vertebrates. Dev Biol 2002; 246(1):45-56.
267. Pöpperl H, Featherstone M. An autoregulatory element of the murine Hox-4.2 gene. EMBO J 1992; 11(10):3673-3680.
268. Gavalas A, Trainor P, Ariza-McNaughton L et al. Synergy between Hoxa1 and Hoxb1: The relationship between arch patterning and the generation of cranial neural crest. Development 2001; 128(15):3017-3027.
269. Arcioni L, Simeone A, Guazzi S et al. The upstream region of the human homeobox gene HOX3D is a target for regulation by retinoic acid and HOX homeoproteins. EMBO J 1992; 11:265-277.
270. Nonchev S, Maconochie M, Gould A et al. Cross-regulatory interactions between Hox genes and the control of segmental expression in the vertebrate central nervous system. Cold Spring Harb Symp Quant Biol 1997; 62:313-323.
271. Yau TO, Kwan CT, Jakt LM et al. Auto/cross-regulation of Hoxb3 expression in posterior hindbrain and spinal cord. Dev Biol 2002; 252(2):287-300.
272. Ferretti E, Cambronero F, Tümpel S et al. The Hoxb1 enhancer and control of rhombomere 4 expression: Complex interplay between PREP1-PBX1-HOXB1 binding sites. Mol Cell Biol 2005; 25:8541-8552.
273. Ferretti E, Marshall H, Pöpperl H et al. Segmental expression of Hoxb2 in r4 requires two separate sites that integrate cooperative interactions between Prep1, Pbx and Hox proteins. Development 2000; 127:155-166.
274. Ferretti E, Schulz H, Talarico D et al. The PBX-regulating protein PREP1 is present in different PBX-complexed forms in mouse. Mech Dev 1999; 83(1-2):53-64.
275. Jacobs Y, Schnabel CA, Cleary ML. Trimeric association of Hox and TALE homeodomain proteins mediates Hoxb2 hindbrain enhancer activity. Mol Cell Biol 1999; 19:5134-5142.
276. Chang CP, Shen WF, Rozenfeld S et al. Pbx proteins display hexapeptide-dependent cooperative DNA binding with a subset of Hox proteins. Genes Dev 1995; 9:663-674.
277. Piper DE, Batchelor AH, Chang CP et al. Structure of a HoxB1-Pbx1 heterodimer bound to DNA: Role of the hexapeptide and a fourth homeodomain helix in complex formation. Cell 1999; 96(4):587-597.
278. Berthelsen J, Zappavigna V, Ferretti E et al. The novel homeoprotein Prep1 modulates Pbx-Hox protein cooperativity. EMBO J 1998; 17(5):1434-1445.
279. Fibi M, Zink B, Kessel M et al. Coding sequence and expression of the homeobox gene Hox 1.3. Development 1988; 102(2):349-359.
280. Deflorian G, Tiso N, Ferretti E et al. Prep1.1 has essential genetic functions in hindbrain development and cranial neural crest cell differentiation. Development 2004; 131(3):613-627.
281. Popperl H, Rikhof H, Chang H et al. Lazarus is a novel pbx gene that globally mediates hox gene function in zebrafish. Mol Cell 2000; 6(2):255-267.
282. Waskiewicz AJ, Rikhof HA, Moens CB. Eliminating zebrafish pbx proteins reveals a hindbrain ground state. Dev Cell 2002; 3(5):723-733.
283. Kimmel CB, Miller CT, Moens CB. Specification and morphogenesis of the zebrafish larval head skeleton. Dev Biol 2001; 233(2):239-257.

Chapter 3

The *Hox* Gene Network in Vertebrate Limb Development

Cheryll Tickle*

Abstract

The *Hox* gene network has multiple roles in vertebrate limb development. One of its main functions is to encode positional information thus providing a "Hox" code for the pattern of structures along the long axis of the limb. Another function of *Hox* genes that has emerged recently is to regulate expression of the *Sonic hedgehog* gene (*Shh*) which controls patterning of distal structures. *Hox* genes also play a major role in development of the digits. The most recent advances have been the identification of control regions that drive *Hox* gene expression in the limb.

Introduction to Vertebrate Limb Development

The first visible sign of vertebrate limb development is the formation of small bulges in the body wall at appropriate positions along the head to tail axis of the embryo. These bulges soon develop into discrete buds, which then grow out substantially (Fig. 1). The buds initially consist of undifferentiated mesenchyme cells encased in ectoderm, but as the bud elongates, cells in the part of the bud nearest the body wall begin to differentiate to form the humerus/femur. The skeleton is then progressively laid down in sequence as the bud grows out with the digits being formed last of all.

Classical embryological experiments in chick limb buds have shown that there are 3 sets of cell-cell interactions in the developing limb bud, one associated with patterning along each of the 3 axes; proximo-distal, dorso-ventral and antero-posterior axes.[1] Laying down the proximo-distal axis of the limb is linked to bud outgrowth which is controlled by the apical ectodermal ridge, the thickened epithelium that rims the tip of the limb bud; dorso-ventral patterning depends on epithelial-mesenchymal interactions between the ectoderm around the sides of the bud and the underlying mesenchyme; and antero-posterior patterning involves signalling by the polarizing region (zone of polarizing activity, ZPA), a region of mesenchyme cells at the posterior margin of the limb bud. When a polarizing region is grafted to the anterior margin of a chick wing bud, anterior cells are respecified to form posterior structures and an additional series of digits develops in mirror-image symmetry with the normal set.[2] Polarizing region signalling also leads to maintenance of the apical ectodermal ridge over the posterior part of the limb bud; in turn signalling by the apical ridge maintains polarizing region signalling.

*Corresponding Author: Cheryll Tickle—Division of Cell and Developmental Biology, School of Life Sciences, University of Dundee, Dow Street, Dundee, DD1 5EH, U.K. Email: c.a.tickle@dundee.ac.uk

HOX Gene Expression, edited by Spyros Papageorgiou. ©2007 Landes Bioscience and Springer Science+Business Media.

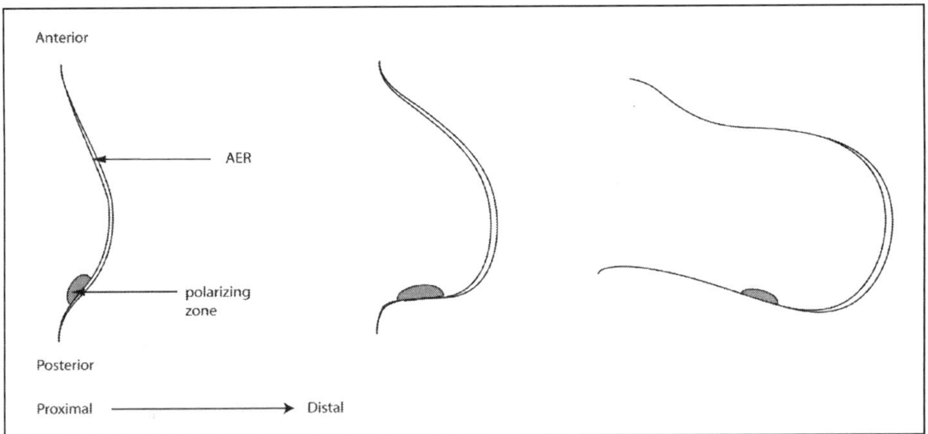

Figure 1. Chick limb development. Diagrams to show stages in development and antero-posterior and proximo-distal axes. The apical ectodermal ridge and polarizing region are also indicated.

Hox Gene Expression in Vertebrate Limbs

Genes in the 5' region of *Hoxa* and *Hoxd* clusters are most prominently expressed in developing vertebrate limbs although some genes in the *Hoxb* and *Hoxc* clusters are also expressed. Indeed, *XlHbox1*, now known as *Hoxc-6,* was one of the first *Hox* genes to be cloned and the first to be shown to be expressed during limb development.[3,4] Staining using specific antibodies showed that Hoxc-6 is expressed in the anterior-proximal region of early forelimb buds in several different vertebrates, including *Xenopus*, zebrafish, mouse and chicken.[3-6] Soon after, Duboule and colleagues showed that 5' *Hoxd* genes (*Hoxd9-13*), then known as *Hox4* genes, are expressed in overlapping domains centred on the posterior-distal margin in both fore and hindlimb buds in mouse embryos.[7,8] There is both temporal and spatial collinearity of expression of these genes in the early limb bud, with 3' genes being expressed before 5' genes and 3' genes being expressed in more proximally than 5' genes (Fig. 2). The expression pattern of each more 5' gene in the cluster is encompassed within the expression pattern of the adjacent 3' gene, and this has been likened to a set of Russian dolls. Similar patterns of *Hoxd* gene expression are also seen in early limb buds in chick embryos and progressive expression of genes in the cluster in a 3' to 5' direction was shown to occur very rapidly within a few hours.[9] 5' *Hoxa* genes were found to be expressed in similar overlapping domains in chick wing buds along the proximo-distal axis but without a clear posterior bias[10] (Fig. 2).

Nelson and colleagues[11] carried out a comprehensive and detailed analysis of expression of all the *Hox* genes in developing chick limbs, including genes from all four clusters. *Hoxc* genes are generally expressed in anterior/proximal regions of either wing or leg bud or both. One cautionary finding was that *Hoxc6* transcripts were detected in both fore and hindlimbs whereas, as previously reported,[6] Hoxc6 protein was only detected in forelimbs. *Hoxb9* is expressed in the anterior of the leg bud and *Hoxb8* expression in the posterior of the early wing bud has subsequently been described by others.[12]

With respect to *Hoxa* and *Hoxd* genes, Nelson et al[11] suggested that that there are at least 3 different phases of expression. In the first phase of expression, *Hoxd9* and *Hoxd10* genes are expressed throughout the lateral plate mesoderm as it begins to thicken to form the bud. Prior to this, a number of different *Hox* genes are expressed in dynamic patterns in lateral plate mesoderm in presumptive limb regions, including other *Hox9* paralogs such as *Hoxb9*. It has been suggested that the expression patterns of these genes at these prelimb stages may serve to position the presumptive limbs along the main head to tail axis of the embryo.[13] The second

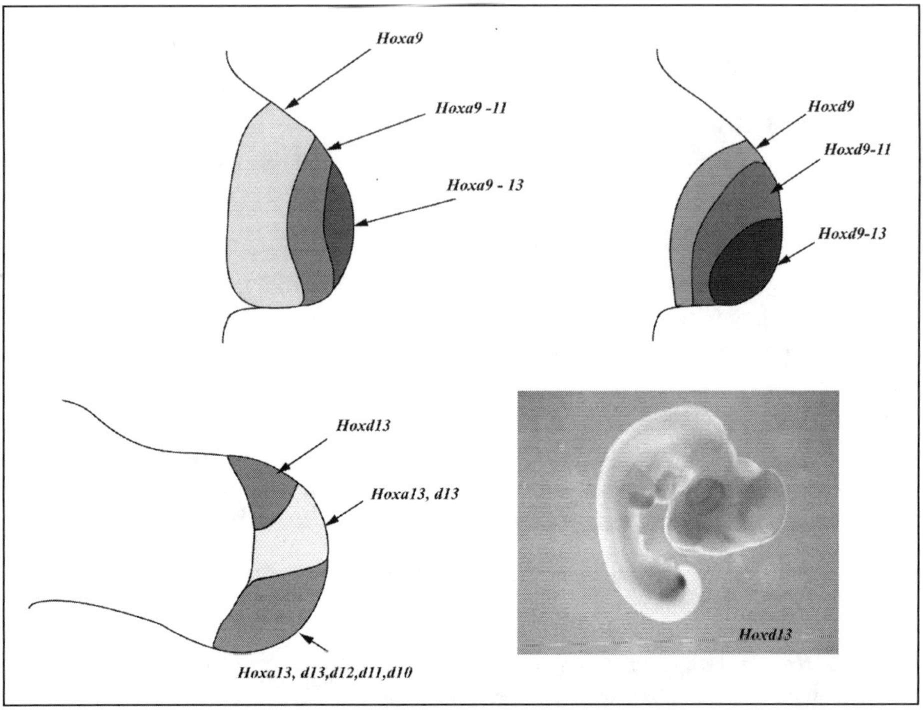

Figure 2. *Hoxd* and *Hoxa* gene expression in developing chick limbs. Top) Diagrams to show nested expression domains of *Hoxa* and *Hoxd* genes in early limb bud, second phase described by Nelson et al.[11] Note that only 3 domains shown for *Hoxd* genes for simplicity, although there are a total of 5 different domains of expression. Below) Later expression of *Hoxa13* and *Hoxd* genes in the region that will form digits, third phase described by Nelson et al[11] and *Hoxd13* expression in the early limb buds of a chick embryo.

phase is the well-described nested patterns of expression already outlined above with *Hoxa13* being expressed later than *Hoxd13*. The third phase of expression is seen in the distal region of more advanced limb buds in the regions where the digits will form (Fig. 2). At this stage, *Hoxd13* is expressed throughout the distal part of the limb and *Hoxd10-Hoxd12* are coexpressed with *Hoxd13* except at the very anterior. *Hoxa13* is also expressed throughout most of the distal region of the limb buds while expression of *Hoxa11* is now restricted more proximally. In this third phase, in addition to the most 5' *Hoxd* genes being expressed in more extensive domains than more 3' genes, they are also expressed temporally in a 5' to 3' direction, with *Hoxd13* being expressed first.

Taken together these data not only reveal specific position-dependent patterns of *Hoxa* and *Hoxd* gene expression in the limb with respect to both antero-posterior and proximo-distal axes but also highlight the dynamic nature of these patterns. The fact that several different phases of expression can be discerned has recently acquired greater significance as control regions that drive *Hox* gene expression in limbs have been uncovered (see later).

Signals That Control *Hox* Gene Expression in the Limb

So what are the signals that lead to the position-dependent patterns of *Hox* gene expression in developing limbs and are these mechanisms consistent with roles for *Hox* genes in limb patterning? These questions were first addressed for Hoxc-6 in chick embryos[6] by carrying out grafting experiments and applying retinoic acid, which was only defined chemical known at that time to be able to mimic polarizing region signalling and produce digit duplications. Since Hoxc6 expression is restricted to the anterior of the early limb bud, one possibility is that Hoxc-6 is involved in encoding antero-posterior positional values. In this case, grafts of the polarizing region or application of retinoic acid to the anterior margin of the wing bud, which respecify anterior cells to form posterior structures would be expected to switch off Hoxc-6 expression in anterior cells. However, instead, these manipulations extended expression and this was associated with shoulder girdle abnormalities. Indeed, abnormalities in this region of the limb skeleton are consistent with a previous fate map which showed that cells in the anterior-proximal region of the limb bud give rise to the shoulder girdle.[14]

The overlapping expression patterns of the 5' *Hoxd* genes in the early bud centred on the posterior-distal margin suggested that these genes might also be involved in antero-posterior limb patterning. *Hoxd13*, for example, is expressed only at the very posterior of the limb bud, while *Hoxd9* is expressed throughout, including the anterior margin (Fig. 2). Indeed, five different expression domains could be distinguished across the antero-posterior axis of the early limb bud leading to the suggestion that this may be why we have five fingers.[15] When polarizing region grafts or retinoic acid-soaked beads were placed at the anterior margin of the chick wing buds, mirror-image patterns of *Hoxd* gene expression resulted and these correlate with the mirror-image patterns of digits that subsequently develop.[9,16] Furthermore, the ectopic domains of expression of the *Hox* genes appeared in the same order as in the normal limb development suggesting that experimental posteriorization of anterior cells recapitulated events that normally occur posteriorly. It was shown that apical ectodermal ridge signalling is also required for induction of ectopic domains of *Hoxd* gene expression in response to retinoic acid. When the apical ridge was removed, the new ectopic domains of *Hox* gene expression at the anterior margin of the wing bud were not induced.[17] The role of the apical ridge in regulating *Hox* gene expression in the limb and the interactions between polarizing region and apical ridge signalling will be discussed later.

It is now known that retinoic acid induces expression of the *Sonic hedgehog* gene, *Shh*.[18] *Shh* is expressed in the polarizing region, and, furthermore, grafting cells expressing Shh to the anterior margin of the chick wing bud can induce mirror-image patterns of *Hox* expression.[18] In mouse mutants with preaxial polydactyly (additional digits at the anterior margin of the limbs), ectopic *Shh* expression has been detected at the anterior of the limb bud and is similarly associated with ectopic *Hoxd* gene expression, see for example.[19,20] In the early chick wing, *Shh* and *Hox* genes are expressed about 24hours after application of retinoic acid to the anterior margin while *Hox* genes are also expressed 24 hours after Shh application. These timings are difficult to reconcile with a simple linear cascade (see later). Furthermore, it has also been shown that Shh can induce mirror-image patterns in late chick limb buds but that, in this case, ectopic *Hoxd13* expression is seen first and in a more extensive domain than more 3' genes.[11]

The mechanism that restricts to 5' *Hoxd* gene expression to posterior-distal mesenchyme in early limb buds is now known to be based on transcriptional repression by Gli3, one of the effectors of Shh signalling. Full length Gli3 protein is processed to a short repressor form in the absence of Shh ligand and there is a gradient of Gli3 repressor in chick and mouse limb buds, with highest levels of repressor in the anterior.[21] In *Shh-/-* mouse embryo limb buds, in which Gli3 repressor will predominate throughout the limb bud, 5' *Hoxd* gene expression in the posterior of the bud is initiated but not maintained and the limbs are severely truncated.[22] In contrast, in limbs of *Gli3-/-* mouse embryos, in which there is no repressor, 5' *Hoxd* genes are uniformly expressed across the antero-posterior axis i.e., in both anterior and posterior mesenchyme.[23,24] Several mouse and chicken mutants have now been discovered with defects in Gli3 processing and, in the limbs of these mutants, 5' *Hoxd* genes are also expressed uniformly across the antero-posterior axis of the limb bud.[21,25-27] In all these mutants, the limbs have many unpatterned digits but it is not clear whether this is directly related to the uniformity of *Hoxd* gene expression.

Bmp2 is another gene that is posteriorly expressed in early chick limb buds and may be repressed by Gli3 in the anterior region of the bud. Bmp signalling has been suggested to mediate the effects of Shh signalling on antero-posterior patterning and digit specification, both in early limb buds and at digital plate stages.[28,29] Application of Bmps to the anterior margin of chick limb buds does not produce mirror-image duplications but instead just an additional digit 2 or a bifurcated digit 3. Nevertheless, changes in 5'*Hoxd* gene expression were seen 48 hours after Bmp application[30] suggesting that Bmps may be upstream of *Hox* gene expression in the digital plate. When the Bmp antagonist, noggin, was used to modulate digit morphogenesis at late stages in development in chick legs, however, no detectable changes in *Hoxd* gene expression were produced.[29]

Apical ridge signalling also plays a role in maintaining *Hox* gene expression in the limb bud. Removal of the apical ridge from early limb buds, which results in limb truncations and lack of distal structures, leads to loss of both *Hoxa13* and *Hoxd13* gene expression.[31-33] Fibroblast Growth Factors mediate apical ridge signalling[34,35] and application of Fgfs to chick wing buds following removal of the apical ridge can rescue both *Hoxd13* and *Hoxa13*.[31,33] These and other data suggest that *Hoxd13* and *Hoxa13* genes could be involved in proximo-distal patterning of the limb and specifically in digit formation.

There has been considerable discussion about how the apical ridge and Fgf signalling is linked to proximo-distal patterning of the limb and laying down the pattern of structures along the long axis of the limb and this is relevant to considering the possible roles of *Hox* genes. A long standing model suggests that proximo-distal patterning is specified by a timing mechanism that operates at the tip of the limb bud in a region of proliferating undifferentiated cells known as the progress zone.[36] A more recent model suggests, in contrast, that all the parts of the limb are already specified in the early bud and that apical ridge signalling then expands these parts sequentially.[37] So how does the behaviour of *Hox* gene expression relate to these different models? The abolition of expression of *Hoxd13* and *Hoxa13* after removal of the apical ectodermal ridge from the early limb bud is consistent with a role for these genes in digit formation and can be explained by both models. In contrast, the establishment of a complete nested set of 5' *Hoxa* and *Hoxd* expression domains in the early bud might fit with the idea of prespecification. Indeed, it has been suggested that the progressive activation of *Hox* gene expression in the early limb bud to establish these nested expression domains could be controlled by a gradient of Fgf, high distally to low proximally[38] emanating from the apical ectodermal ridge. However, application of Fgf to chick wing buds did not lead to premature expression of *Hoxa* genes suggesting that other factors are involved.[33] On the other hand, the behaviour of cells expressing *Hoxd13* is difficult to square with the idea of prespecification. Cell marking experiments have indicated that, as the limb bud grows out, some cells that expressed *Hoxd13* in the early limb bud become displaced from the limb bud tip and cease to express *Hoxd13*.[39] One interpretation of these data is that distal cells become progressively proximalised during limb bud outgrowth.

The general conclusion from the data above is that manipulations of the limb that produce changes in pattern formation are accompanied by changes in *Hox* gene expression and that the cell-cell signalling molecules involved in patterning, such as Shh and Fgfs are involved in maintaining the *Hox* gene expression network. However, it is worth noting that the signals that mediate patterning and limb bud outgrowth are coordinated. Thus, there is a positive feedback loop in which Shh maintains expression of various *Fgf* genes in the apical ectodermal ridge, which in turn maintain *Shh* expression in the polarizing region.[40,41] Therefore changing one signal can have knock-on effects. This makes it difficult to interpret such experiments and to separate the roles of *Hox* genes in antero-posterior versus proximo-distal patterning.

Tests of *Hox* Gene Function

Two main approaches have been used to test directly the function of *Hox* genes in vertebrate limb development—misexpression or overexpression and functional inactivation. The function of

5' *Hoxd* genes has been examined using both approaches. In the first approach in chick embryos, *Hoxd11* (previously known as *Hox4.6*) was over-expressed in developing limbs using the RCAS (replication competent avian sarcoma virus) system. In these experiments, an additional digit 2 was produced in the wing, while, in the leg, digit 1 appeared to be transformed into a digit II, ie posteriorised.[42] Chicken *Hoxd12* has been ectopically expressed in mouse embryos using the Hoxb6 promoter, which drives expression in lateral plate mesoderm and throughout hind-limb bud mesenchyme.[43] This misexpression of *Hoxd12* also resulted in posteriorization of the limbs, including conversion of digit I to the morphology typical of digit II and/or duplications of anterior digits. Although the initial chick embryo experiments were interpreted as being due to a direct effect of *Hoxd11*, it was shown in the mouse limbs, in which *Hoxd12* was misexpressed, that there is ectopic *Shh* expression at the anterior margin. This led to the suggestion that *Hoxd12,* and possibly other 5'*Hox* genes might play a role in reinforcing *Shh* expression at the posterior margin of the growing limb (see later).

In the second approach, transgenic mice have been created in which *Hox* genes have been functionally inactivated. In the first experiments, single *Hoxd* genes, such as *Hoxd13* were knocked out.[44] The limbs of *Hoxd13-/-* mice showed complex changes in digit anatomy, including both reduction in the size of the digits, missing phalanges and fused digits but also additional digits in the forelimbs. It is not easy to interpret these changes in anatomy simply in terms of the hypothesis that *Hoxd* genes encode digit identities across the antero-posterior axis of the limb. They are however consistent for a role for *Hoxd13* in the formation of distal structures. Furthermore, mice with a mutation in *Hoxa11* showed defects in ulna/radius and tibia/fibula regions,[45] again pointing to a role in proximo-distal patterning, but, in other single mutants, such as those in which *Hoxd11* was functionally inactivated, the hindlimbs appeared relatively normal.

It is now clear that there are compensatory and synergistic interactions both between paralogous *Hox* genes in different clusters and between *Hox* genes within the same cluster and that these interactions affect the phenotype. Thus when both *Hoxa13* and *Hoxd13* were functionally inactivated, the mice lacked digits altogether.[46] In other experiments, knock-outs have been made in which more than one 5' *Hoxd* gene have been functionally inactivated and these experiments show that several different paralogs contribute to digit patterning. Thus, for example, in trans-heterozygotes of *Hoxd13* and *Hoxd12*, novel digital abnormalities were obtained and the defects seen in individual heterozygotes were exacerbated,[47] while simultaneous deletion of *Hoxd13*, *Hoxd12* and *Hoxd11* resulted in reduced number and size of digits compared to deletion of *Hoxd13* alone.[48] A series of compound mutants with varying degrees of loss function of *Hoxd13*, *Hoxa13* and *Hoxd11/12* have also been genetically engineered.[49] Analysis of the digit phenotypes of these mutants suggest that it is the dose of *Hox* genes that is important for digital development rather than qualitative differences between genes. Interestingly, the progressive deletion of *Hox* gene function generated, first of all, limbs with extra digits, and then limbs with reduced numbers of digits.

The limb phenotypes of these and other compound *Hox* gene mutant mice have recently been reviewed[50] and a 'Hox' code for the limb has been deduced (Fig. 3). This code functions primarily with respect to proximo-distal limb pattern. According to this code, *Hox13* paralogs function predominantly in the digits, *Hox11* in radius/ulna and tibia/fibula, and *Hox9* in the humerus and *Hox10* in the femur. However some of these genes also play a role in other parts of the limb. For example, *Hox11* paralogs also contribute to digit patterning and *Hoxd12* can substitute for *Hox13* paralogs. One of the biggest outstanding challenges is to identify downstream target genes that translate the *Hox* code into anatomy. A recent report provides evidence that an ephrin receptor, *EphA7,* is a target of Hoxd13 and Hoxa13.[51]

An even more drastic elimination of *Hox* gene expression in mouse limb buds has been achieved recently and confirms the importance of the *Hox* genes in proximo-distal outgrowth and patterning.[52] Genes in both the *Hoxd* and *Hoxa* clusters have been conditionally knocked-out in the mesenchyme of the limb buds using *Prx1-Cre* mice. *Prx1-Cre* is more efficient in the

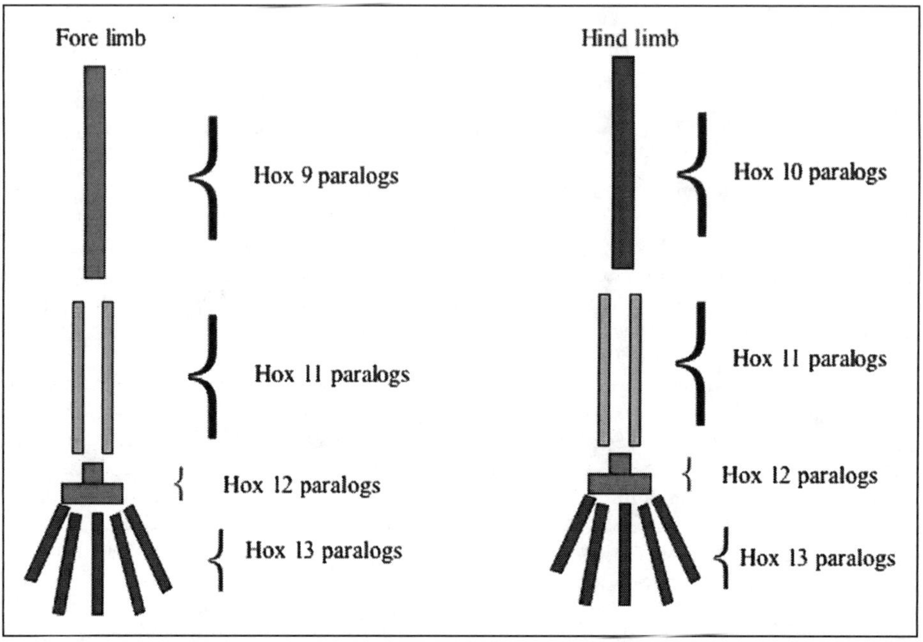

Figure 3. "Hox" code for limb development, after Wellik and Capecchi.[50]

forelimbs and therefore the phenotype of the forelimbs was reported. The forelimbs of the conditional double-mutants were extremely truncated and distal structures completely missing. Interestingly, this loss of distal structures was associated with lack of *Shh* expression in the limb buds. The phenotype is however more severe than that of the limbs of *Shh-/-* mouse embryos in which there are remnants of *Hox* gene expression.

Other *Hox* genes may also be involved in establishing *Shh* expression in limb buds. *Hoxb8* is expressed at the very posterior of early mouse limb buds. An unexpected role for *Hoxb8* was uncovered when *Hoxb8* was expressed more anteriorly using the RAR beta2 promoter.[53] This resulted in polydactylous forelimbs with ectopic expression of *Shh* at the anterior margin of the forelimb. Experiments in chick embryos also supported a role for *Hoxb8* gene in establishing *Shh* expression in the polarizing region of the forelimb and showed that *Hoxb8* expression can be regulated by retinoic acid.[12,54] Recently it has been reported that a microRNA, miR-196, may provide an inhibitory mechanism that prevents *Hoxb8* being expressed in the hindlimb.[55]

Regulatory Regions That Drive *Hox* Gene Expression in the Limb

Considerable progress has been made in identifying control regions in the DNA that drive *Hoxd* gene expression in the limb through sophisticated experiments in mice (Fig. 4). One global control region (GCR) lies 5' to the cluster and directs expression in the digital region of the limb,[56] equivalent to the third phase of expression described by Nelson et al.[11] Interestingly, in this region, there are other genes, *Lunapark* and *Evx2,* which are expressed in the digital region in the same way as *Hoxd* genes but may have little function, if any, in digit development. A DNA segment was identified in the mouse that can drive expression of a LacZ reporter in the digital region of the limb and in the neural tube.

The existence of another control region (ECLR), this time driving *Hoxd* cluster gene expression in early limbs, has been defined through experiments in which an inversion and large

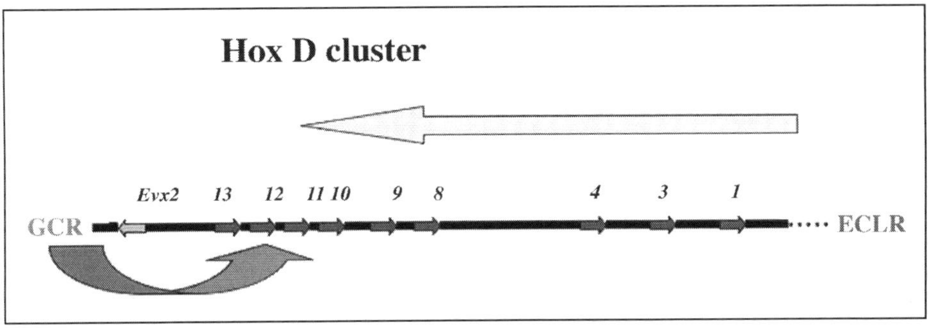

Figure 4. Control regions of the *HoxD* cluster, after Zakany et al.[57]

deficiency in the mouse cluster was engineered.[57] These manipulations resulted in 5' *Hoxd* genes being expressed in the pattern of 3' *Hoxd* genes within early limb bud, such that *Hoxd13* expression came on first and was expressed throughout the early bud, instead of being posteriorly restricted. This led to the development of polydactylous limbs consistent with the previous overexpression studies in chick and mouse.[42,43] Furthermore, like the mouse experiments, anterior expression of 5'genes was associated with ectopic expression of *Shh* and suggests a role for *Hox* genes in establishing *Shh* expression in the polarizing region. As already mentioned, this conclusion has since been supported by the analysis of the conditional double-mutants for *Hoxa* and *Hoxd* genes.[52] These experiments further suggest that there is a control region that lies 3' to the *HoxD* cluster that directs an early phase of expression in the limb bud, equivalent to second phase described by Nelson et al.[11] More recent work has confirmed the idea that two control regions serve to drive early and late phases of *Hoxd* gene expression in the limb, and that these two phases are concerned with development of proximal structures and distal structures respectively.[58]

Wider Implications of the *Hox* Gene Network

The substantial efforts that have been directed towards understanding the fundamental role of the *Hox* genes in limb development are now reaping rewards. The focus on *Hox* genes has been particularly relevant to human limb malformations. The first two limb malformations found to be due to mutations in *Hox* genes were hand-foot-genital syndrome[59] and synpolydactyly[60] with the gene affected in each case being *Hoxa13* and *Hoxd13* respectively. Other mutations in *Hoxd13* can give rise to different phenotypes such as brachydactyly associated with central polydactyly.[61,62] Ultimately, one would like to explain how such specific gene mutations lead to such precise changes in anatomy.

Comparative studies have also led to some intriguing speculations about vertebrate limb evolution. In embryonic fin buds of both a teleost (zebrafish)[63] and a paddlefish,[64] *Hoxa13* and *Hoxa11* are coexpressed distally at late stages. Furthermore, the third phase of *Hoxd* gene expression appears to be absent in both pectoral and pelvic fin buds of zebrafish.[63] More recently, the regulatory regions that drive digital *Hoxd* gene expression have been compared across species by phylogenetic fingerprinting. The DNA segment found in the mouse genome that drives LacZ expression in the digits and neural tube was found to be highly conserved in both human and pufferfish genomes. Intriguingly, when the equivalent pufferfish sequence was used to drive LacZ in mouse embryos, only the neural tube expression was obtained. This suggests that a tetrapod digit enhancer may be missing from the teleost DNA segment. This is a very provocative finding, given the suggestion that the digits are an evolutionary invention of higher vertebrates. It seems certain that a deeper understanding of the *Hox* gene network will lead to new insights into vertebrate limb origins.

Acknowledgments

I thank Angie Blake for help with the preparation of the manuscript, Andy Bain and Eva Tiecke for making the figures, and the MRC and The Royal Society for funding my research.

References

1. Saunders JW. The experimental analysis of chick limb bud development. In: Ede DA H, JR, Balls MJ, eds. Vertebrate Limb and Somite Morphogenesis. Cambridge: Cambridge University Press, 1977:1-24.
2. Saunders JW, Gasseling MT. Ectodermal-mesenchymal interactions in the origin of limb symmetry. In: Fleischmeyer R, Billingham RE, eds. Epithelial-mesenchymal interactions. Baltimore: Williams and Wilkins, 1968:78-97.
3. Oliver G, Wright CV, Hardwicke J et al. A gradient of homeodomain protein in developing forelimbs of Xenopus and mouse embryos. Cell 1988; 55(6):1017-1024.
4. Oliver G, Wright CV, Hardwicke J et al. Differential antero-posterior expression of two proteins encoded by a homeobox gene in Xenopus and mouse embryos. EMBO J 1988; 7(10):3199-3209.
5. Molven A, Wright CV, Bremiller R et al. Expression of a homeobox gene product in normal and mutant zebrafish embryos: Evolution of the tetrapod body plan. Development 1990; 109(2):279-288.
6. Oliver G, De Robertis EM, Wolpert L et al. Expression of a homeobox gene in the chick wing bud following application of retinoic acid and grafts of polarizing region tissue. EMBO J 1990; 9(10):3093-3099.
7. Dolle P, Duboule D. Two gene members of the murine HOX-5 complex show regional and cell-type specific expression in developing limbs and gonads. EMBO J 1989; 8(5):1507-1515.
8. Dolle P, Izpisua-Belmonte JC, Falkenstein H et al. Coordinate expression of the murine Hox-5 complex homeobox-containing genes during limb pattern formation. Nature 1989; 342(6251):767-772.
9. Izpisua-Belmonte JC, Tickle C, Dolle P et al. Expression of the homeobox Hox-4 genes and the specification of position in chick wing development. Nature 1991; 350(6319):585-589.
10. Yokouchi Y, Sasaki H, Kuroiwa A. Homeobox gene expression correlated with the bifurcation process of limb cartilage development. Nature 1991; 353 (6343):443-445.
11. Nelson CE, Morgan BA, Burke AC et al. Analysis of Hox gene expression in the chick limb bud. Development 1996; 122(5):1449-1466.
12. Stratford TH, Kostakopoulou K, Maden M. Hoxb-8 has a role in establishing early anterior-posterior polarity in chick forelimb but not hindlimb. Development 1997; 124(21):4225-4234.
13. Cohn MJ, Patel K, Krumlauf R et al. Hox9 genes and vertebrate limb specification. Nature 1997; 387(6628):97-101.
14. Saunders JW. The proximo-distal sequence of origin of parts of the chick wing and the role of the ectoderm. Journal of Experimental Zoology 1948; 108:363-404.
15. Tabin CJ. Why we have (only) five fingers per hand: Hox genes and evolution of paired limbs. Development 1992; 116(2):289-296.
16. Nohno T, Noji S, Koyama E et al. Involvement of the Chox-4 chicken homeobox genes in determination of anteroposterior axial polarity during limb development. Cell 1991; 64(6):1197-1205.
17. Izpisua-Belmonte JC, Brown JM, Duboule D et al. Expression of Hox-4 genes in the chick wing links pattern formation to the epithelial-mesenchymal interactions that mediate growth. EMBO J 1992; 11(4):1451-1457.
18. Riddle RD, Johnson RL, Laufer E et al. Sonic hedgehog mediates the polarizing activity of the ZPA. Cell 1993; 75(7):1401-1416.
19. Masuya H, Sagai T, Moriwaki K et al. Multigenic control of the localization of the zone of polarizing activity in limb morphogenesis in the mouse. Dev Biol 1997; 182(1):42-51.
20. Sharpe J, Lettice L, Hecksher-Sorensen J et al. Identification of sonic hedgehog as a candidate gene responsible for the polydactylous mouse mutant Sasquatch. Curr Biol 1999; 9(2):97-100.
21. Wang B, Fallon JF, Beachy PA. Hedgehog-regulated processing of Gli3 produces an anterior/posterior repressor gradient in the developing vertebrate limb. Cell 2000; 100(4):423-434.
22. Chiang C, Litingtung Y, Harris MP, et al. Manifestation of the limb prepattern: Limb development in the absence of sonic hedgehog function. Dev Biol 2001; 236(2):421-435.
23. Litingtung Y, Dahn RD, Li YN et al. Shh and Gli3 are dispensable for limb skeleton formation but regulate digit number and identity. Nature 2002; 418(6901):979-983.
24. te Welscher P, Zuniga A, Kuijper S et al. Progression of vertebrate limb development through SHH-mediated counteraction of GLI3. Science 2002; 298(5594):827-830.
25. Liu A, Wang B, Niswander LA. Mouse intraflagellar transport proteins regulate both the activator and repressor functions of Gli transcription factors. Development 2005; 132(13):3103-3111.

26. Caruccio NC, Martinez-Lopez A, Harris M et al. Constitutive activation of sonic hedgehog signaling in the chicken mutant talpid(2): Shh-independent outgrowth and polarizing activity. Dev Biol 1999; 212 9(1):137-149.
27. Ispizua-Belmonte JC, Ede DA, Tickle C et al. The mis-expression of posterior Hox-4 genes in talpid (ta3) mutant wings correlates with absence of antero-posterior polarity. Development 1992; 114(4):959-963.
28. Yang Y, Drossopoulou G, Chuang PT et al. Relationship between dose, distance and time in Sonic Hedgehog-mediated regulation of anteroposterior polarity in the chick limb. Development 1997; 124(21):4393-4404.
29. Dahn RD, Fallon JF. Interdigital regulation of digit identity and homeotic transformation by modulated BMP signaling. Science 2000; 289(5478):438-441.
30. Duprez DM, Kostakopoulou K, Francis-West PH et al. Activation of Fgf-4 and HoxD gene expression by BMP-2 expressing cells in the developing chick limb. Development 1996; 122(6):1821-1828.
31. Vogel A, Roberts-Clarke D, Niswander L. Effect of FGF on gene expression in chick limb bud cells in vivo and in vitro. Dev Biol 1995; 171(2):507-520.
32. Hashimoto K, Yokouchi Y, Yamamoto M et al. Distinct signalling molecules control Hox-a11 and Hox-a13 expression in the muscle precursor and mesenchyme of the chick limb bud. Development 1999; 126(12):2771-2783.
33. Vargesson N, Kostakopoulou K, Drossopoulou G et al. Characterisation of Hoxa gene expression in the chick limb bud in response to FGF. Dev Dyn 2001; 220(1):87-90.
34. Niswander L, Tickle C, Vogel A et al. FGF-4 replaces the apical ectodermal ridge and directs outgrowth and patterning of the limb. Cell 1993; 75(3):579-587.
35. Fallon JF, Lopez A, Ros MA et al. FGF-2: Apical ectodermal ridge growth signal for chick limb development. Science 1994; 264(5155):104-107.
36. Summerbell D, Lewis J, Wolpert L. Positional information in chick limb morphogenesis. Nature 1973; 244(5413):492-496.
37. Dudley AT, Ros MA, Tabin CJ. A reexamination of proximodistal patterning during vertebrate limb development. Nature 2002; 418(6897):539-544.
38. Papageorgiou S. Cooperating morphogens control hoxd gene expression in the developing vertebrate limb. J Theor Biol 1998; 192(1):43-53.
39. Vargesson N, Clarke JD, Vincent K et al. Cell fate in the chick limb bud and relationship to gene expression. Development 1997; 124(10):1909-1918.
40. Niswander L, Jeffrey S, Martin GR et al. A positive feedback loop coordinates growth and patterning in the vertebrate limb. Nature 1994; 371(6498):609-612.
41. Laufer E, Nelson CE, Johnson RL et al. Sonic hedgehog and Fgf-4 act through a signaling cascade and feedback loop to integrate growth and patterning of the developing limb bud. Cell 1994; 79(6):993-1003.
42. Morgan BA, Izpisua-Belmonte JC, Duboule D et al. Targeted misexpression of Hox-4.6 in the avian limb bud causes apparent homeotic transformations. Nature 1992; 358(6383):236-239.
43. Knezevic V, De Santo R, Schughart K et al. Hoxd-12 differentially affects preaxial and postaxial chondrogenic branches in the limb and regulates Sonic hedgehog in a positive feedback loop. Development 1997; 124(22):4523-4536.
44. Dolle P, Dierich A, LeMeur M et al. Disruption of the Hoxd-13 gene induces localized heterochrony leading to mice with neotenic limbs. Cell 1993; 75(3):431-441.
45. Small KM, Potter SS. Homeotic transformations and limb defects in Hox A11 mutant mice. Genes and Development 1993; 7(12A):2318-2328.
46. Fromental-Ramain C, Warot X, Messadecq N et al. Hoxa13 and Hoxd13 play a crucial role in the patterning of the limb autopod. Development 1996; 122:2997-3011.
47. Davis AP, Capecchi MR. A mutational analysis of the 5' HoxD genes: Dissection of genetic interactions during limb development in the mouse. Development 1996; 122(4):1175-1185.
48. Zakany J, Duboule D. Synpolydactyly in mice with a targeted deficiency in the HoxD complex. Nature 1996; 384(6604):69-71.
49. Zakany J, Fromental-Ramain C, Warot X et al. Regulation of number and size of digits by posterior Hox genes: A dose-dependent mechanism with potential evolutionary implications. Proc Natl Acad Sci USA 1997; 94(25):13695-13700.
50. Wellik DM, Capecchi MR. Hox10 and Hox11 genes are required to globally pattern the mammalian skeleton. Science 2003; 301(5631):363-367.
51. Salsi V, Zappavigna V. Hoxd13 and Hoxa13 directly control the expression of the EphA7 Ephrin tyrosine kinase receptor in developing limbs. J Biol Chem 2006; 281(4):1992-1999.

52. Kmita M, Tarchini B, Zakany J et al. Early developmental arrest of mammalian limbs lacking HoxA/HoxD gene function. Nature 2005; 435(7045):1113-1116.
53. Charite J, de Graaff W, Shen S et al. Ectopic expression of Hoxb-8 causes duplication of the ZPA in the forelimb and homeotic transformation of axial structures. Cell 1994; 78(4):589-601.
54. Lu HC, Revelli JP, Goering L et al. Retinoid signalling is required for the establishment of a ZPA and for the expression of Hoxb-8, a mediator of ZPA formation. Development 1997; 124(9):1643-1651.
55. Hornstein E, Mansfield JH, Yekta S et al. The microRNA miR-196 acts upstream of Hoxb8 and Shh in limb development. Nature 2005; 438(7068):671-674.
56. Spitz F, Gonzalez F, Duboule D. A Global control region defines a chromosomal regulatory landscape containing the HoxD cluster. Cell 2003; 113(3):405-417.
57. Zakany J, Kmita M, Duboule D. A dual role for Hox genes in limb anterior-posterior asymmetry. Science 2004; 304(5677):1669-1672.
58. Tarchini B, Duboule D. Control of Hoxd genes' collinearity during early limb development. Dev Cell 2006; 10(1):93-103.
59. Mortlock DP, Innis JW. Mutation of HOXA13 in hand-foot-genital syndrome. Nat Genet 1997; 15(2):179-180.
60. Muragaki Y, Mundlos S, Upton J et al. Altered growth and branching patterns in synpolydactyly caused by mutations in HOXD13. Science 1996; 272(5261):548-551.
61. Goodman FR. Limb malformations and the human HOX genes. Am J Med Genet 2002; 112(3):256-265.
62. Caronia G, Goodman FR, McKeown CM et al. An I47L substitution in the HOXD13 homeodomain causes a novel human limb malformation by producing a selective loss of function. Development 2003; 130(8):1701-1712.
63. Sordino P, van der Hoeven F, Duboule D. Hox gene expression in teleost fins and the origin of vertebrate digits. Nature 1995; 375(6533):678-681.
64. Metscher BD, Takahashi K, Crow K et al. Expression of Hoxa-11 and Hoxa-13 in the pectoral fin of a basal ray-finnned fish, Polydon spathula: Implications for the origin of tetrapod limbs. Evol Dev 2005; 7(3):186-195.

CHAPTER 4

Evolution of Hox Gene Clusters

David E.K. Ferrier*

Introduction

The Hox gene clusters have been one of the most prominent paradigms within Developmental Biology. This stems from the great excitement that surrounded the discovery that the genes all contained the conserved homeobox motif and that the homologous genes were operating in broadly homologous ways in the development of organisms as phylogenetically widespread as flies and vertebrates. The sequence similarity between the genes based on the homeobox, and their specific genomic organization in both flies and vertebrates, immediately implied a particular mode of evolution of the Hox gene cluster by tandem duplication and, more intriguingly, a functional constraint on the organization of the cluster to conserve colinearity. This general picture still holds true, but our understanding of the nature and extent of the constraints on cluster organization have been modified in recent years as data has become available from a much wider selection of animal phyla.

Colinearity, whereby the order of the genes along the chromosome corresponds to (or is colinear with) the domains of action of the genes along the anterior-posterior axis, has been one of the most intriguing and mysterious phenomena associated with the Hox cluster. Despite intense effort we do not have a clear idea about its mechanistic basis. It has even been argued that there may not be a universal underlying mechanism.[1] Whilst this probable lack of a universal mechanism operating across the various different instances of colinearity in an animal like the mouse is becoming clearer, whether this 'complicated' view can be extended to a lack of a universal, ancestral mechanism of colinearity (with add-on, lineage-specific mechanisms elaborating the picture) is unresolved. We must adopt a broad, phylogenetically informed, comparative approach (incorporating genomics, phylogenetics and developmental biology) to have a chance of finding an ancestral mechanism and finally unraveling the mystery of colinearity.

Origin of the Cluster: Unraveling Patterns of Gene Duplication and Duplicate Evolution

Homeobox Phylogenies, Tandem Duplication and the ProtoHox Cluster

Through the intricate and insightful genetic work of Ed Lewis the Bithorax Complex (BX-C) portion of the *Drosophila melanogaster* Hox cluster was elucidated. Lewis developed the idea that the gene complex consisted of a series of duplicated elements, or pseudoalleles as he first called them.[2-4] In time the BX-C was found to consist of three genes,[5] and the homeotic complex for the anterior end of the fly, the Antennapedia Complex (ANT-C), was found to contain 5 homeotic genes with a handful of interspersed nonhomeotic loci.[6] In the 1980's the homeobox was discovered in the genes of these complexes,[7] and this 180bp motif rapidly

*Correspondence: David E.K. Ferrier—Department of Zoology, University of Oxford, Tinbergen Building, South Parks Road, Oxford, OX1 3PS, U.K. Email: david.ferrier@zoo.ox.ac.uk

HOX Gene Expression, edited by Spyros Papageorgiou. ©2007 Landes Bioscience and Springer Science+Business Media.

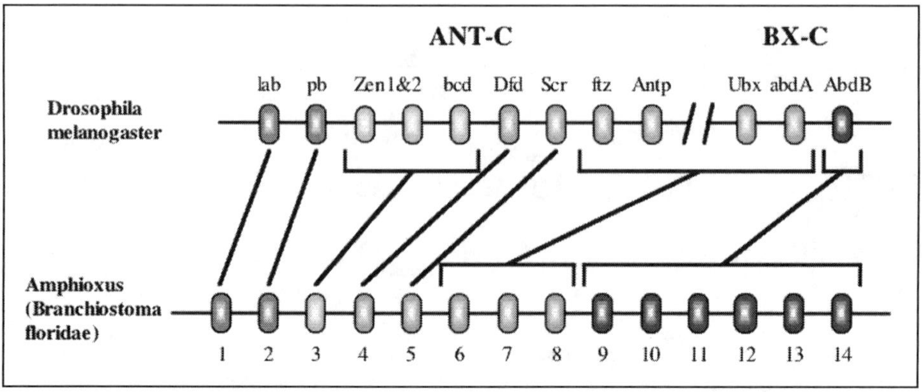

Figure 1. The homeobox gene content of the Hox clusters of *Drosophila melanogaster* and amphioxus. The fly cluster is broken into two, the ANT-C and BX-C. The ANT-C contains some homeobox genes that are derived from Hox genes but have evolved new, nonHox functions (*zen 1, zen2, bcd* and *ftz*). *Zen 1, zen2* and *bcd* evolved from duplications of an ancestral *Hox3* gene.[83,84] Several of the Central (green) genes cannot be distinguished as direct orthologues between insects and chordates, and at least some of them may have evolved from independent sets of tandem duplications. The single Posterior Hox gene (dark blue), *AbdB*, in flies is orthologous to the multiple Posterior genes of chordates, *Hox9-14*. The amphioxus cluster is prototypical for the multiple clusters of vertebrates (e.g., the four mammalian Hox clusters HoxA, B, C and D), which arose at the origin of the vertebrates, with subsequent gene loss occurring within each vertebrate cluster. A color version of this figure is available online at www.Eurekah.com.

found in the genomes of other animals, including ourselves. Together the ANT-C and BX-C of flies are homologous to the Hox clusters of chordates (Fig. 1).

The widespread conservation of the homeobox and its presence in a large family of transcription factor-encoding genes enabled the rapid isolation of multitudes of homeobox genes from a diverse array of phyla. Construction of phylogenetic trees of homeodomain sequences has been vital in understanding their modes of evolution and their classification.[8] From such trees it is evident that the Hox genes are closely related to each other, that is, the cluster is not an amalgamation of disparate, unrelated genes. Clearly this is consistent with the Hox cluster having formed via tandem duplication (Fig. 2). The specific nature of these tandem duplications has been debated. Some have proposed that the 'multiplication' of the genes in the cluster has occurred via unequal cross-overs, such that the genes inside the cluster are 'chimaeras' of the flanking genes.[9] Alternatively tandem duplication without chimaeric gene formation is also clearly possible. Indeed there are instances of this having occurred in the evolution of some present-day Hox clusters. For example the *Zerknüllt* (*zen*) genes of *Drosophila* are duplicated and are clearly related to each other rather than being chimaeras of flanking genes.

In addition to the homeobox phylogenies revealing the close relationship of the Hox genes to each other, it is also clear that several other genes that are not members of Hox clusters are interspersed amongst the Hox genes themselves in these trees. It was supposed that these represented dispersed or orphan Hox genes that had evolved by trans duplication of particular Hox genes deep in animal evolution, so that some Hox-related genes became scattered around the genome. This view was transformed when the ParaHox cluster was discovered in the cephalochordate amphioxus (*Branchiostoma floridae*).[10] The three genes of the ParaHox cluster are Gsx, Xlox and Cdx (in humans GSH1, IPF1 and CDX2). In homeobox phylogenies Gsx groups with the anterior Hox genes of groups 1 and 2, Xlox groups with Hox3 genes, and Cdx groups with the Posterior Hox genes. These gene relationships and the order of Gsx, Xlox and Cdx along the chromosome are consistent with a model in which the Hox and ParaHox clusters arose from a common ancestral homeobox cluster by duplication (the ProtoHox hypothesis).[10-14]

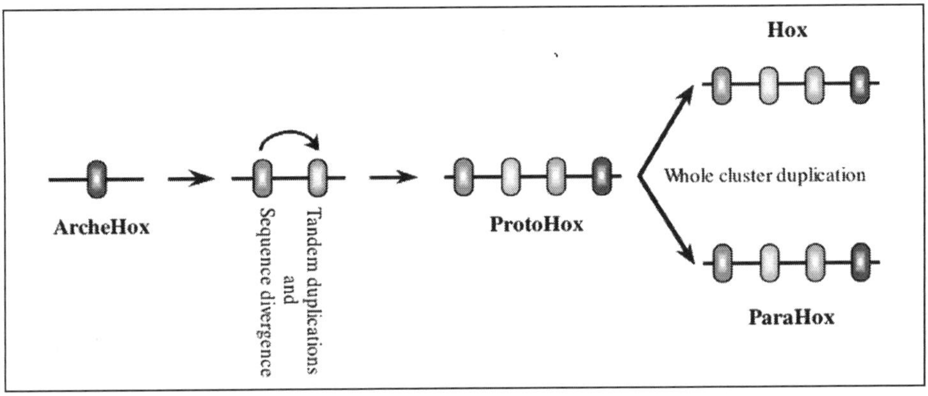

Figure 2. The origin of the Hox genes. An ancestral Hox-like gene, ArcheHox, underwent a series of tandem duplications and the duplicated genes diverged to establish the precursors of each of the different classes of Hox gene, Anterior (red), Group 3 (pale blue), Central (green) and Posterior (dark blue), in the ProtoHox cluster. The entire ProtoHox cluster duplicated, and gave rise to the Hox cluster and its evolutionary sister the ParaHox cluster. The precise timing of these duplication events is uncertain, as is whether the ProtoHox cluster contained a precursor of each of the 4 main groups of Hox gene or not (see the text). A color version of this figure is available online at www Eurekah.com.

Furthermore the expression of the ParaHox genes also exhibits Colinearity, with Gsx being expressed from the anterior of the embryo/larva, Xlox in mid-body regions and Cdx at the posterior end of the organism.[10] The ParaHox cluster is thus the evolutionary sister, or paralogue, of the Hox cluster (Fig. 2).

Basal Animals and the "Trox2" Model of Duplicate Evolution

When did this transition from a ProtoHox cluster to the Hox and ParaHox clusters occur? The answer is that it was clearly very early in the evolution of animals, but the precise time relative to the origin of particular animal phyla is still to be resolved. When the ParaHox cluster was first discovered it was thought that the ProtoHox duplication may have occurred at the transition between the radially symmetrical, diploblastic animals and the triploblastic bilaterians.[10] However as more sequences have become available from diploblasts, such as the cnidarians, it has become clear that both Hox and ParaHox genes are present in this diploblastic phylum.[11-13] So we must look deeper into the animal phylogeny to find the transition. However this becomes problematic due to the lack of clear consensus on the relationships of these basal animal phyla (Porifera, Placozoa, Cnidaria and Ctenophora).[15-17] Further work on the phylogeny of these diploblast phyla is required to clearly resolve their relationships and the ordering of their divergences from the lineage leading to the Bilateria, so that we can relate the homeobox content of the phyla to models of cluster evolution.

The Porifera, or sponges, are one candidate for the most basal animal group.[15,16] No clear Hox-like or ProtoHox genes have been found so far, although the search continues and will be aided by whole genome sequences of sponges. The enigmatic diploblast phylum Placozoa is an alternative candidate as the most basal animal phylum[17] (Fig. 3), and so may also provide us with a crucial data point in understanding the origin of the Hox cluster.

Trox2 is the only Hox-like gene to have been found in *Trichoplax adhaerens*, a placozoan. This gene however raises as many issues as it answers. Thorough searches have been performed for ANTP-class genes in *Trichoplax* (A.S. Monteiro et al in press), and no other Hox-like genes besides *Trox2* have been found. So potentially *Trox2* could be directly descended from an ancestral ProtoHox gene, or even the ArcheHox gene. However the sequence of Trox2 does not behave as expected for a ProtoHox protein in phylogenetic trees. Trox2 groups robustly with

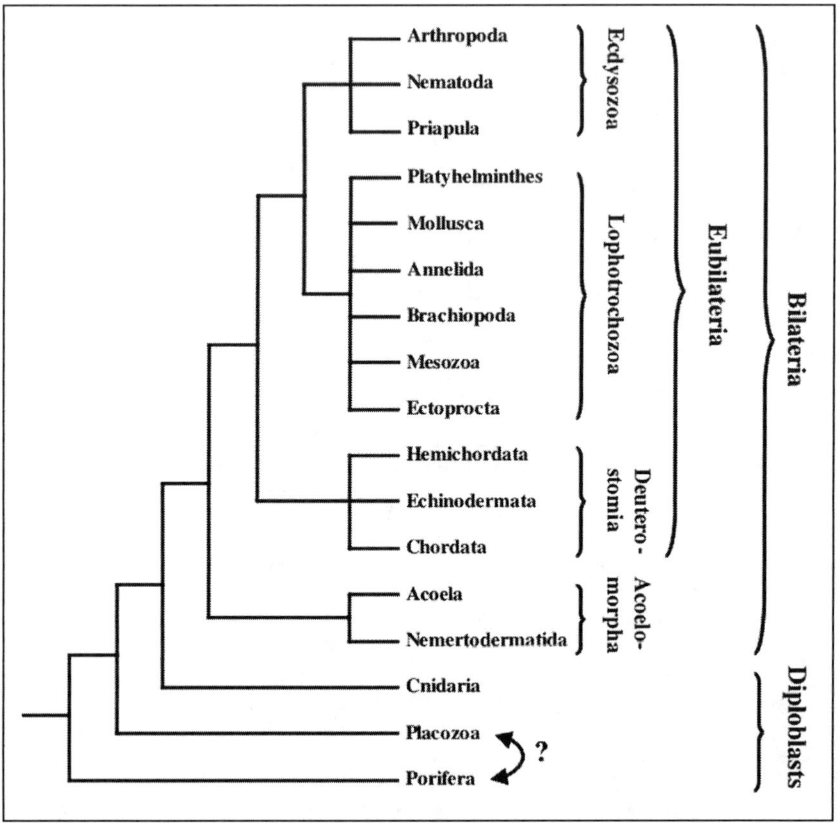

Figure 3. A phylogeny of the animal kingdom. The tree represents a consensus of references 21, 29, 32, 35, 36 and 85. The most basal animal lineage is not well resolved.[15-17]

the ParaHox protein Gsx.[18-20] So is *Trox2* really a descendent of a ProtoHox gene, or is it a Gsx gene and other Hox and ParaHox genes have been lost in the *Trichoplax* lineage? It is very difficult to distinguish between these two possibilities. However by building phylogenies of other *Trichoplax* genes and seeing how they relate to their bilaterian homologues we can see if many other *Trichoplax* genes behave as does *Trox2*, i.e., grouping with particular bilaterian genes within the bilaterian gene families rather than corroborating the a priori expectation that the *Trichoplax* gene would be basal or sister to the entire bilaterian gene family (Fig. 4A). If further *Trichoplax* genes behave like *Trox2* in phylogenetic trees then this could be taken as evidence for extensive gene loss in placozoans, which has hit the homeobox genes just as much as other gene families.

Alternatively *Trichoplax* genes are descended from ancestral 'Proto' genes and our models and assumptions about sequence evolution following gene duplication need to be modified. If the *Trichoplax* genes reflect the ancestral condition, and *Trox2* is a direct descendent from an ArcheHox/ProtoHox gene without duplication for example, then rather than both daughters of a gene duplication event diverging from the ancestral, preduplicate sequence, only one daughter diverges whilst the other retains the characteristics of the ancestral sequence (Fig. 4B). This revolutionary view of gene evolution may seem reasonable if the ancestral Proto gene is embedded in the developmental networks of the animal, with all of the consequent constraints on its function and hence sequence. After the duplication event all of these constraints are still present, and they are all still focused on one of the daughters, if for

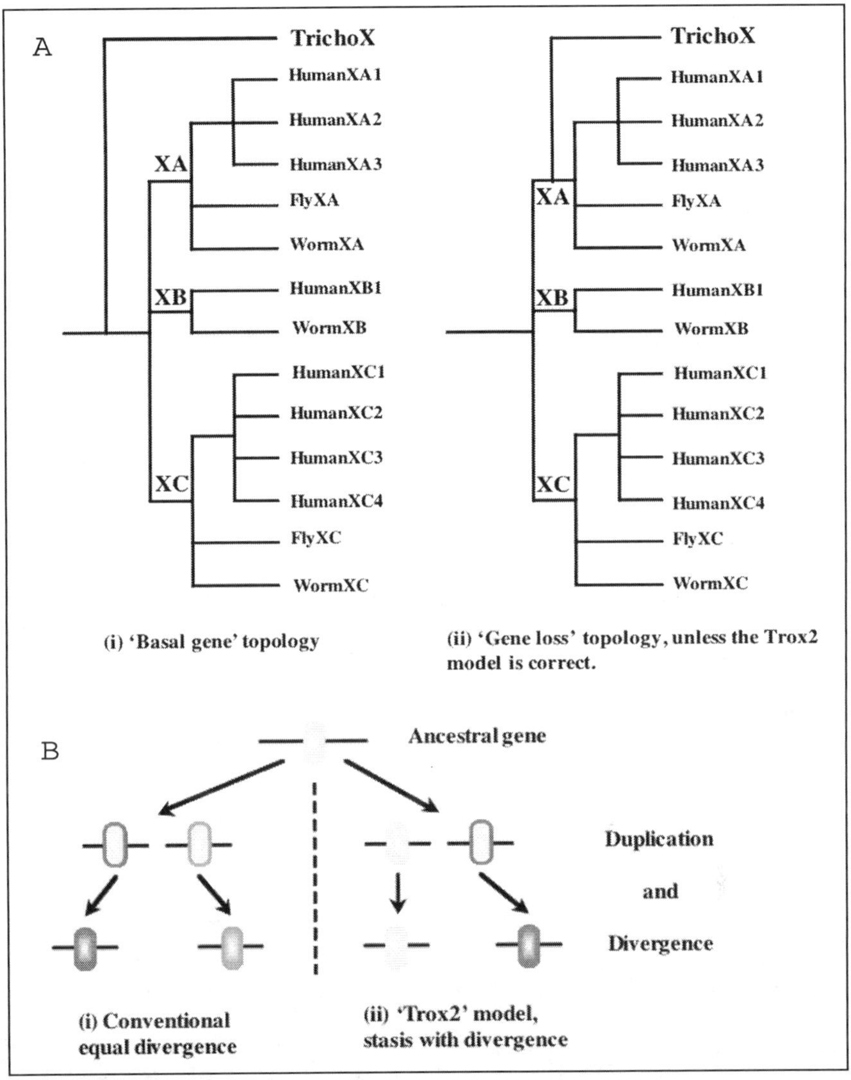

Figure 4. A) Inference of ancestral gene content from molecular phylogeny topologies. A gene family (X) has a diversity of members in the Bilateria (XA, XB and XC), whilst the placozoan *Trichoplax* has a single X gene (TrichoX). (i) If the Trichoplax gene is a direct, unduplicated descendent of the gene ancestral to XA, XB and XC, then TrichoX should diverge from a node more basal than the separations of the XA, XB and XC groups. (ii) If TrichoX groups with a particular bilaterian group such as XA with a node more crownward than the separation of XA, XB and XC, then we can infer that TrichoX is an XA gene and Trichoplax has lost its XB and XC genes. If in fact TrichoX really is a direct, unduplicated descendent of the ancestral X gene but still groups robustly with a particular clade such as XA, then an unconventional mode of gene evolution, the 'Trox2' model, has occurred. B) The Trox2 model of gene duplicate evolution. (i) Conventionally it is assumed that after a gene duplication event both daughter genes will diverge from the ancestral sequence. (ii) An alternative pattern of sequence evolution is the one that has been proposed to have occurred for Trox2,[86] whereby one of the duplication daughters diverges from the ancestral sequence, whilst the other daughter remains largely unchanged. A color version of this figure is available online at www.Eurekah.com.

example only one of the daughters has retained all of the regulatory elements and hence the full expression pattern of the ancestral gene, whilst the other daughter consists of the full coding sequence but not all of the regulatory 'baggage'. The result is that the second daughter is less constrained, diverges from the ancestral sequence and then settles into a new developmental network before it becomes a nonfunctional pseudogene. This stabilises its sequence so that it is conserved and recognisable across subsequently evolving phyla, and is recognised as a new, distinct gene class.

Perhaps whole genome sequences from basal animal groups such as Porifera, Placozoa and Cnidaria will help to resolve the issue of ancestral Hox-like gene content, and hopefully organization, and whether this rather revolutionary view of duplicated gene evolution is real or not. However due to the extensive time that has elapsed, combined with lineage-specific genomic evolution (e.g., the cnidarian Hox-like genes may even be independent duplications from a ProtoHox condition relative to the bilaterian Hox/ParaHox genes) even whole genome sequences may well not help, as no extant animal may have retained enough of the ancestral condition for us to discover. An alternative approach that must be pursued in parallel is to sample more widespread bilaterian taxa to establish patterns of conservation, and extrapolate back into the animal phylogeny to reconstruct ancestral conditions.

ProtoHox Content and the Nature of its Duplication

This extrapolation approach was how the ProtoHox hypothesis was first constructed, with the Hox cluster originating from the duplication of a 4-gene ProtoHox cluster. Alternative models of 2- and 3-gene ProtoHox clusters have since also been proposed.[12-14] The 4-gene ProtoHox model stems from the fact that Gsx groups with Hox1-2, Xlox with Hox3 and Cdx with Posterior Hox genes. There is not a ParaHox gene in the chordate ParaHox clusters that is allied with the Central Hox genes (groups 4-8). The 4-gene model thus posits that this fourth ParaHox gene was lost after the duplication of the ProtoHox cluster. An equally parsimonious scenario is a 3-gene ProtoHox model in which the Central Hox genes evolved after the ProtoHox duplication, so that the ParaHox cluster only ever had Gsx, Xlox and Cdx.[12,13] More recently a 2-gene ProtoHox model has been developed,[14,19-22] which proposes that Xlox, in the ParaHox cluster, and Hox3, in the Hox cluster, did not have a common ancestral gene in the ProtoHox cluster, but evolved by independent duplication in their respective clusters followed by convergence of their sequences, so that their grouping in phylogenetic trees is no longer a true reflection of their ancestry. In this 2-gene, as with the 3-gene scenario, the Central Hox genes then evolved just in the Hox cluster well after the ProtoHox duplication (reviewed in refs. 14,22).

The genomic organization of the Hox cluster has been one of its most prominent features, given its link to the functioning of the genes to produce Colinearity. The genomic organization around the cluster has also been informative, not necessarily with relevance to Hox function, but more with regard to the mode of the evolution of the cluster. Evx is another homeobox gene within the Antp-superclass that lies within a group of genes called the Extended Hox class,[23,24] not only because of its location in phylogenetic trees, but also because in several organisms it is tightly linked to the Hox cluster. This linkage clearly reflects the ancestral condition, and is still found in chordates and cnidarians.[25-27] Gauchat et al[19] proposed that Evx and the ProtoHox/ArcheHox gene arose from a duplication event. Minguillón and Garcia-Fernàndez[24] on the other hand propose that it was not Evx that was produced by this first duplication that gave rise to the initial ProtoHox gene, but rather the ancestor of both Evx and Mox was produced. Mox is another member of the Extended Hox class, and is also linked to the Hox clusters of vertebrates.[23,24] Mox and Evx genes show a weak association in phylogenetic trees. Taken in concert with their linkage to the vertebrate Hox clusters, and nonhomeobox genes that neighbour the Hox and ParaHox clusters, it appears that the duplication of the ProtoHox cluster may have been a tandem event, and ancestrally the newly formed Hox and ParaHox clusters were adjacent to each other, only subsequently being separated to different regions of the genome by translocation.[24]

Evolution of Cluster Composition and the Impact of Hox on Animal Phylogeny

The Key Position of the Acoelomorpha

Although there is the uncertainty about the exact composition of the ProtoHox cluster and where gene losses or gains happened in relation to the origin of particular diploblast phyla, as outlined above, the pattern of four basic groups of Hox genes (anterior, group3, central, posterior) had been established by the origin of the Eubilateria. Whether the full complement of the four Hox groups was also present right at the origin of the basal Bilateria is not completely resolved at present (see Fig. 3 and Baguñà and Riutort[21] for a discussion of the distinction between Bilateria and Eubilateria), and to a certain extent is linked to which of the 3/4-gene models or the 2-gene model of the ProtoHox cluster is true.

A phylogeny of the animal kingdom is of course essential for our interpretations of Hox evolution. A present consensus would be something like (Fig. 3). The phylogeny of the diploblasts is uncertain at present, but in recent years major advances have been made in resolving the tree of the Bilateria. The Acoelomorpha (= Acoela + Nemertodermatida) are likely to be basal bilaterians, and as such can provide us with an important outgroup to the higher bilaterians (eubilaterians) (reviewed in ref. 21). The Acoela have Hox genes that fall into the Anterior, Central and Posterior groups, but no Hox3 group has yet been found.[28] In the other Acoelomorph group, the Nemertodermatida, a fragment of an Xlox ParaHox gene has been found (Eva Jimenez-Guri pers. comm.), in addition to posterior and central Hox genes. If the 3 or 4-gene models of the ProtoHox are true, then the presence of an Xlox in Nemertodermatids means that the Xlox/Hox3 group was present in the Last Common Bilaterian (LCB). If the 2-gene ProtoHox model reflects reality then we still do not know whether the Hox3 group was in the LCB, or only Xlox had originated before the LCB, and Hox3 did not appear until after the divergence of the Acoelomorpha lineage but before the Last Common Eubilaterian (LCE).

Characteristic Hox Content of the Ecdysozoa, Lophotrochozoa and Deuterostomia

The Eubilaterians are conventionally divided into the protostomes and deuterostomes, and the protostomes in turn split into some variant of the Ecdysozoa and Lophotrochozoa.[29-31] Distinctive patterns of Hox genes seem to be broadly representative for each of these major eubilaterian groups.[32] In particular the Central Hox genes seem to be characteristic for each group, possibly reflecting independent duplications and expansions of this region of the Hox cluster in each of the three eubilaterian groups.[33] The types of Posterior Hox genes are even more clearly distinctive for each eubilaterian group,[32] with Ecdysozoa having variants of AbdB, Lophotrochozoa having Post1 and Post2, and deuterostomes having multiple Posterior Hox genes that are given the names Hox9-Hox14[34] (although this deuterostome nomenclature encompasses some extensive sequence diversity (see below)).

Such distinctive patterns of Hox gene possession have been useful in determining the phylogenetic position of several enigmatic phyla: Mesozoa,[35] Bryozoa/Ectoprocta,[36] and Brachiopoda,[32] although using Hox genes in such a fashion must be done with care.[37] One enigmatic group that could well benefit from further investigations of its Hox gene content to resolve its phylogenetic position is the Chaetognatha. This phylum has been allied with all three of the eubilaterian groups at one time or another: deuterostomes due to their embryology, Ecdysozoa due to their 18S rDNA sequences, and most recently the Lophotrochozoa via their mitochondrial sequences (mtDNA).[38] The difficulty in locating the Chaetognatha is exemplified by the same form of data, the mtDNA, being analysed independently and producing slightly different answers. In both analyses of the mtDNA the Chaetognatha group with the protostomes, but in one analysis they are basal protostomes and in the other they are lophotrochozoans.[38-40] Some information is available on chaetognath Hox genes. Unfortunately there is insufficient sequence information from the Central Hox genes of chaetognaths to determine which genes are present as yet (Lox5/Lox2/Lox4 or Antp/abdA/Ubx, or Hox6/7/8), and no Posterior sequences have been cloned so far. However one gene in chaetognaths, *SceMedPost,* is distinctive.[41]

SceMedPost was proposed to be a mosaic gene, sharing sequence characteristics between Medial/Central and Posterior Hox genes.[41] This was hypothesised to reflect an ancestral state prior to divergence into Medial and Posterior states, with the chaetognath lineage originating prior to the split of deuterostomes and protostomes. The latest molecular data suggests that this basal bilaterian position of the chaetognaths is not true, and that they are in fact with the protostomes. Also it is now clear that the distinction between Posterior and Central/Medial Hox genes had already occurred at the base of the bilaterians, from the Acoelomorph data.[28] *SceMedPost* may yet provide us with some phylogenetic clues however. Another Hox gene that has always proved difficult to classify as Medial or Posterior is the *egl-5* gene of nematodes. Originally *egl-5* was thought to be the Posterior Hox gene of *Caenorhabditis elegans* due to its position at the end of the nematode Hox cluster (although the use of the term 'cluster' can only be loosely applied to nematodes). Other, clearer Posterior Hox genes have since been found in nematodes,[42] and egl-5 never groups robustly with Posterior Hox proteins in phylogenetic trees (e.g., see ref. 32). Intriguingly some of the residues that are considered to provide 'Medial' characteristics to SceMedPost are also present in egl-5 (namely Q6, T7, I45 and E59). Might this then indicate a phylogenetic affinity of chaetognaths and nematodes? This would be consistent with some phylogenetic reconstructions,[16,43] and the possession of some Ecdysozoan characters by chaetognaths, such as a ventral nerve chord combined with radial cleavage. Importantly however chaetognaths are thought not to undergo ecdysis, which led Peterson and Eernisse[16] to speculate that chaetognaths may be basal Ecdysozoans. This loose similarity of SceMedPost and nematode egl5 is clearly weak evidence on its own, but sequences of Posterior Hox genes may well be much more revealing if chaetognaths are found to have AbdB/nob1/php3 sequences.

How Many Posterior Hox Genes Did the Ancestors Have?

The Posterior Hox genes are evidently extremely useful tools for resolution of the broad relationships of animal phyla, however our understanding of the evolution of these Hox genes themselves is far from complete. Even the basic question as to how many Posterior Hox genes did the bilaterian ancestor have is far from resolved. Historically the Hox community has usually assumed that there was a single Posterior Hox, and the multiple genes seen in vertebrates arose from a series of tandem duplications in the chordate or deuterostome lineage. Such a view was inevitably coloured by the first known Hox cluster being that of *Drosophila,* and flies only having a single Posterior Hox gene, *Abdominal-B* (*AbdB*). If we examine the Posterior Hox gene content of a broader range of phyla in the context of a phylogenetic tree, then we can immediately see that the possession of a single Posterior Hox gene in flies is rather unusual.[32] Even the most closely related phyla to the arthropods (i.e., other Ecdysozoans) for which we have some data (nematodes and priapulids) seem to have more than one Posterior Hox gene (*nob1* and *php3* in *C.elegans* and *Pca-AbdB* and *Pca-HB4* in *Priapulus caudatus*).[32,42] Elsewhere, in the Lophotrochozoa, there are at least two Posterior Hox genes (Post1 and Post2), and in deuterostomes there are a whole array of Posterior Hox genes. Chordates can have up to six (Hox9-Hox14),[34,44] whilst the full complement of more basal deuterostomes (echinoderms and hemichordates) is still to be determined. This lack of resolution of the basal deuterostome condition is present even though we have a completely cloned and sequenced Hox cluster from an echinoderm, the purple sea urchin *Stronglyocentrotus purpuratus.*[45] Contrary to first impressions of the urchin Hox cluster[46] the organization of the genes is extremely derived and scrambled.[45] Gene loss has also evidently occurred relative to the ancestral echinoderm condition, as Hox4 has been lost in *S.purpuratus* but is present in asteroids,[47] and other echinoderms have a more extensive set of Posterior Hox genes than the purple sea urchin.[48,49]

The relationships between the Posterior Hox genes of chordates and those of Ambulacraria (echinoderms and hemichordates) is also difficult to resolve, which leaves uncertainty as to what the Posterior Hox content of the ancestral deuterostome actually was. Although vertebrates, amphioxus and urochordates are all said to have a Hox10 for example, the genes of this name from each of these three groups of chordates do not actually group together robustly in phylogenetic trees. This lack of resolution can be extended to the nonchordate deuterostomes such as the sea urchin, which has a so-called Hox9/10 gene, the name arising because of its loose association with the chordate Hox9 and Hox10 groups but its lack of clear resolution with either particular group. Such a lack of resolution of the orthology relationships amongst the deuterostome Posterior Hox genes is in stark contrast to the Posterior genes of protostomes, in which robust groupings of AbdB, Post1 and Post2 are formed even between phyla. This contrasting behaviour of the Posterior Hox genes in phylogenetic trees was given the name Deuterostome Posterior Flexibility.[44] It was hypothesised to be the result of higher rates of sequence evolution in the deuterostome Posterior Hox genes than in those of protostomes, which in turn leads to a lack of resolution of the deuterostome gene relationships when constructing trees. However since the original Deuterostome Posterior Flexibility hypothesis was proposed more gene sequences have been isolated. In particular some Posterior Hox genes have been cloned from a hemichordate, and a couple of these acorn worm sequences can clearly be identified as orthologues of some echinoderm counterparts.[50] This led to the suggestion that Posterior Flexibility is only a chordate phenomenon rather than a deuterostome-wide mode of evolution.[50] Several of the hemichordate Posterior Hox genes still do not resolve clearly with echinoderm counterparts however, and so one alternative scenario that has been proposed is that the Posterior Hox genes of chordates and Ambulacraria are the result of two independent sets of tandem duplications.[45] The patterns of the deuterostome Posterior Hox gene groupings in phylogenetic trees however do not show two independent sets of Posterior Hox genes, one in Ambulacraria and one in chordates. Consequently independent duplications do not seem to be a plausible sole explanation. Perhaps the deuterostome Posterior Hox genes have evolved via a mixture of the two phenomena, lineage-specific duplications and higher rates of evolution as outlined by the Deuterostome Posterior Flexibility hypothesis. Further deuterostome Posterior Hox sequences, accompanied by careful phylogenetic tree building, may help to resolve the issue. One intriguing possibility is that an extant basal deuterostome, *Xenoturbella*,[51] may be available to provide another perspective on deuterostome Posterior Hox gene evolution.

It is of course very difficult to confidently resolve such issues with a handful of gene fragments from a few distinct phyla dotted around the animal kingdom. In the not-too-distant future we should have greater sampling of a more extensive diversity of taxa, and entire Hox clusters cloned and sequenced. Such entire cluster sequences will greatly improve our understanding by providing positional information within a cluster, and more importantly providing us with the entire Hox gene complement for the relevant taxon. For example does the chaetognath MedPost gene exist alongside Posterior genes, specifically Post1 and Post2 or AbdB? Are the acoelomorph Hox genes clustered, and are there other genes that have been missed by screens so far (e.g., more Acoel Central genes and Hox3)? What is the basal Hox complement for the Ambulacraria? What Hox genes are present in *Xenoturbella*, and can this animal help us to determine what the basal deuterostome condition was with regards to Hox cluster composition?

So the Hox cluster is a veritable maelstrom of gene duplications and losses across the Bilateria, and there are many more examples than those outlined above. As we sample the Hox genes across a broader phylogenetic sample of taxa we will have a much clearer picture of the evolution of Hox cluster composition, and will also probably discover more Hox characters that can resolve debates about various issues in animal phylogenetics.

Molecular Mechanisms and Evolution of Hox Cluster Organization

The Importance of Temporal Colinearity

Colinearity in the Hox cluster can take different forms; spatial, temporal, or quantitative.[1] To a certain extent the three types of Colinearity are not always mechanistically independent. For example Temporal Colinearity may be a route to Spatial Colinearity in some circumstances.[52] However we can find instances of one form of Colinearity occurring in the absence of any others, e.g., the axial Spatial Colinearity of the *Drosophila* Hox gene expression without any obvious Temporal or Quantitative components.

Myself and others have hypothesised that it is Temporal Colinearity that is the key to understanding Hox cluster organization.[52-57] This hypothesis is mainly based upon the observation that Hox clusters that conform to our view of the ancestral condition, of an ordered cluster of genes with anterior expression beginning at one end followed by a gradual progression through to posterior expression at the other end of the intact, complete cluster, is found in animals and clusters exhibiting Temporal Colinearity. In those taxa which do not or cannot use a temporal component in the initiation of their Hox expression, then the Hox cluster tends to be broken, dispersed and rearranged. Spatial Colinearity does not require an intact cluster, as is evident from flies, nematodes and urochordates,[57] and so the mechanistic basis for Spatial Colinearity seems an unlikely means for understanding the maintenance and organization of the Hox cluster.

Broken, dispersed Hox clusters are found in taxa with rapid modes of development, which also often correlates with a relatively low number of cells in the embryo.[54] Such a mode of development is generally considered to be rather derived within each respective lineage, such as the insects or chordates, which ancestrally developed gradually by progressive elongation of the posterior end of the embryo.[58] A more recent example of another broken, derived Hox cluster is that of the Schistosome flatworm, *Schistosoma mansoni*.[59] Again this correlates with a derived mode of development and life-style within the lineage, as Schistosomes are extremely specialised parasites, with a highly specialised and derived life-cycle to match.

When the hypothesis that Temporal Colinearity is the main constraining force on Hox cluster organization was formulated[53,54] the echinoderm Hox cluster was apparently an exception to the rule, due to its intact, well-ordered nature but with a clear lack of Temporal Colinearity.[46,60] With the recent clarification of the organization of the *S.purpuratus* Hox cluster we now know that the urchin Hox cluster is highly derived after all,[45] which correlates with its extreme form of indirect development, with almost complete loss of embryonic structures at metamorphosis, and a highly derived adult form that cannot easily be compared to the morphology of other phyla. The echinoderm Hox genes are however still maintained as a cluster and not dispersed like the clusters of flies, nematodes, schistosomes and urochordates. Such cluster maintenance may be indicative of enhancer sharing amongst the echinoderm Hox genes, which is largely unaffected by reordering of the genes but does require them to remain in close proximity.

The clearest evidence for an importance of time in the control of Hox expression comes from work in mammals. Duboule has pointed out that colinearity is obeyed most rigorously at the time of initial expression.[61] Investigations of enhancers of Hox genes, using LacZ reporters, show that those that seemingly reproduce the complete expression pattern of the relevant Hox gene still have one aberration from the gene itself. The timing of their initiation is later than the endogenous gene.[1,62] Also mutant phenotypes are more severe in animals that have lost only a subset of the genes from a cluster, compared to animals that have lost an entire cluster; deletion of the entire HoxC cluster of mice does not produce pronounced homeotic mutants.[63] Crawford[55] proposes that the deletion of a subset of Hox genes alters the timing and progress of gene initiation through the Hox cluster, leading to more severe homeotic phenotypes than whole cluster deletion. In a sense the experiment has also been done naturally. Teleosts have undergone an extra genome duplication relative to the tetrapods, but only have seven Hox clusters rather than the expected eight, due to the loss of an entire HoxD or HoxC cluster.[64] That the mammalian Hox clusters are gradually unwound and de-repressed during the earliest stages of development is now clear.[65,66] The trigger for this process, and how it links to the activating enhancers of the Hox genes will be an important revelation. There is a caveat to this scenario so far however. Translocation of an Anterior Hox gene to a location near HoxD13 in mice does not cause the Anterior Hox gene to now be expressed at the time of a Posterior gene. It is still activated early, at least in the mesoderm.[1,67] It would be intriguing to know how the chromatin organization changes during the activation of this modified Hox cluster.

Alternatively Bilaterian Hox Colinearity Mechanisms Are not Homologous

A flip-side to these considerations of the mechanistic basis for Hox cluster integrity is to question whether the cluster really is as constrained as we thought it was? Could clustering simply be an indication of evolutionary history by tandem duplications of genes, followed by subsequent different break-ups in divergent lineages? An immediate riposte would be, why have linkages been conserved for so long? Perhaps different constraints were added on in different lineages after the initial gene origins by tandem duplications, i.e., is Colinearity in protostomes mechanistically comparable to Colinearity in deuterostomes, or even can comparisons be made within deuterostomes themselves? Duboule and coworkers have shown that in vertebrates there are several mechanisms contributing to Hox Colinearity, depending on the context. Thus there is no single universal mechanism of Colinearity in vertebrate Hox clusters. But this diversity of mechanisms must have been imposed on vertebrate Hox clusters, evolving from an ancestral cluster that presumably already exhibited Colinearity, perhaps resembling the Colinearity seen in the amphioxus Hox cluster.[68] Can we determine what the mechanistic basis for this ancestral Colinearity was? Is it a mechanism that is still used in vertebrates, presumably in a context that was also present in the ancestor, e.g., body axis or CNS patterning, rather than a vertebrate innovation such as limb patterning, and is this mechanism present outside of the deuterostomes?

The alternative hypothetical scenario is thus that the genes originated by tandem duplication, they use shared enhancers due to these tandem duplications, and this enhancer-sharing reduces the opportunity for viable genomic rearrangements of the cluster so that it is conserved for longer than would be a cluster of genes without enhancer-sharing. This cluster maintenance perhaps only needs to be kept for a relatively short time (in geological terms), maybe between the late Vendian Ediacaran animals and the explosive divergence of the bilaterian lineages in the Cambrian. At this point different Colinearity mechanisms then evolve along some Hox clusters of different animal lineages, whilst in other lineages the cluster finally disperses as rare, viable genomic rearrangements (with the appropriate enhancers) slowly accumulate (Fig. 5). Sharing of enhancers between more than one Hox gene occurs in mice and flies, and enhancers are widespread and densely packed throughout the Hox clusters.[69-71] This would fit with the above scenario, whereby recombination within the cluster would be deleterious more often than not, and so slows cluster disintegration considerably. A test of this hypothesis requires comparison of the organization of Hox clusters in more taxa, to see how often Hox clusters have disintegrated, followed by the characterization of the regulatory mechanisms to find the degree of mechanistic conservation across phyla.

Furthermore we can form an estimate of the time that it might take for cluster break-up to occur when there is enhancer-sharing but no global Colinearity mechanism. Importantly this time is longer than the Ediacara-Cambrian explosion period. The flies seem to provide us with a group of organisms that do not have a global Colinearity mechanism operating across their Hox cluster, since the cluster is breaking up in the Drosophilids. The Hox clusters of several species of *Drosophila* have been sequenced, and breaks have occurred in different locations within the cluster.[72-75] These rearrangements can be linked to the estimates for divergence times amongst *Drosophila* lineages, and it can be seen that only three different cluster breaks have occurred along lineages that have been separated for 30-60 Million Years.[76] The presence of broken Hox clusters also extends more deeply into the insects,[77] and so potentially the release from the constraints on clustering is even more ancient than the origin of the Drosophilids, and our estimate of three breaks in three lineages of 30-60 million years is rather conservative.

If the fossil record is to be believed, and the Ediacaran faunas represent the early stages of animal evolution with the diversification of the bilaterian lineages not happening until the Cambrian, then a period of less than 20 Million Years covers the origin of the ArcheHox gene (sometime after the origin of animals), the expansion to the ProtoHox cluster, the duplication into the Hox and ParaHox clusters, and then finally the origin of the bilaterians and their subsequent explosive radiation. Consequently the time that elapsed between the origin of the Hox cluster and the divergence of the bilaterian lineages may well be only a few million years, and a shorter period of time than has elapsed since the divergence of the various *Drosophila* lineages, in which no global Colinearity mechanism exists and yet still only a few viable Hox cluster breaks have evolved. It may thus be perfectly plausible that the ancestral Hox cluster was conserved as a cluster, by enhancer-sharing for example, without the constraint of a global Colinearity mechanism until the Bilateria diverged, and then lineage-specific Hox Colinearity mechanisms evolved after the bilaterian divergence (Fig. 5ii).

It has been postulated that two of the best-studied model organisms with regards to Hox gene function (namely *Drosophila* and mice) control and use their Hox genes in fundamentally different ways—flies are 'qualitative' whilst mice tend to be 'quantitative'.[1] Perhaps we cannot see the mechanistic connections and commonalities because we are dealing with the two extremes of a continuum. Other taxa must be studied, with less specialised, derived modes of development than the fly, and less redundancy than the mouse (i.e., a single cluster instead of four). This may well help us to penetrate the evolutionary fog accumulated around mice and flies. Despite the extreme situation with flies and mice we do still have some mechanistic starting points from fly-mouse comparisons. Both use Polycomb/Trithorax group genes to regulate their Hox genes,[78] and both have boundary or insulator elements in their Hox clusters,[79,80] as well as microRNAs.[81] Do other taxa? Are these Hox-specific mechanisms, or tools used by other clusters of genes (Polycomb group complexes are widespread across the chromosomes,[78] and interactions of protein-coding genes with microRNAs are common)?[82] Are they integral to the (ancestral) mechanisms of Colinearity? There are many open questions, and we are at a very early stage in understanding Hox Colinearity.

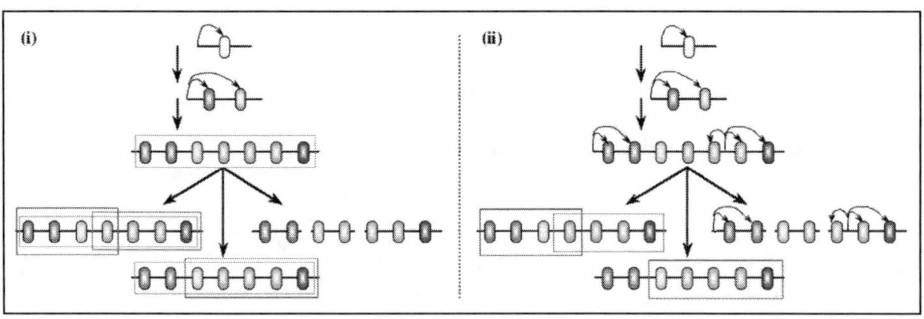

Figure 5. Is there a homologous mechanism of Colinearity amongst different bilaterian lineages? The Hox cluster evolved by tandem duplications from an ancestral ArcheHox gene (yellow). Evolution by tandem duplication could well have resulted in enhancer-sharing by multiple genes (curved arrows). In scenario (i) a Colinear mechanism (orange box) evolved before the divergence of the different bilaterian lineages. In the separate lineages further lineage-specific Colinear mechanisms evolved (blue, purple and brown boxes), and in some lineages the ancestral Colinear mechanism is lost and the cluster subsequently broken after the removal of this evolutionary constraint (discontinuous horizontal lines). In scenario (ii) the prevalence of enhancer-sharing in the ancestral Hox cluster results in viable breaks in the cluster being very unlikely. Consequently the cluster is maintained until after the divergence of the bilaterian lineages. Subsequently lineage-specific Colinearity mechanisms evolve in some lineages, whilst in others sufficient time eventually elapses for viable cluster breaks to evolve. A color version of this figure is available online at www.Eurekah.com.

Conclusion

From the earliest molecular days of the Hox cluster some of the fundamental elements of the evolution of the cluster were known, e.g., tandem duplication and Colinearity. But in the intervening years significant shifts have occurred: origin from a ProtoHox cluster, extensive cluster disintegration, distinct patterns of gene evolution across the cluster (such as Posterior versus Anterior), and between groups of animals, e.g., Ecdysozoa, Lophotrochozoa, Deuterostomia. Looking to the future many fundamental questions remain, such as the ancestral composition of the cluster (for animals as a whole, for basal bilaterians and eubilaterians, and for each of the major clades of Ecdysozoa, Lophotrochozoa and Deuterostomia), when did the ProtoHox to Hox/ParaHox transition occur, are there general cluster-wide mechanisms constraining the cluster across the animals, or is the conservation of clusters mechanistically different in separate lineages (if so was there an ancestral mechanism or does the cluster simply reflect the mode of gene evolution by tandem duplication)? Hox cluster sequencing from a greater diversity of taxa, combined with gene expression work in the light of the organization of the relevant clusters, and ultimately elucidation of gene regulation mechanisms in a diversity of taxa, will hopefully one day show us how and why the Hox cluster exists.

References

1. Kmita M, Duboule D. Organizing Axes in time and space; 25 years of colinear tinkering. Science 2003; 301:331-333.
2. Lewis EB. Pseudoallelism and gene evolution. Cold Spring Harb Symp Quant Biol 1951; 16:159-174.
3. Lewis EB. Genes and developmental pathways. Am Zoologist 1963; 3:33-56.
4. Lewis EB. A gene complex controlling segmentation in Drosophila. Nature 1978; 276:565-570.
5. Sánchez-Herrero E, Vernós I, Marco R et al. Genetic organization of Drosophila bithorax complex. Nature 1985; 313:108-113.
6. Kaufman TC, Lewis R, Wakimoto B. Cytogenetic analysis of chromosome 3 in Drosophila melanogaster: The homeotic gene complex in polytene chromosome interval 84A-B. Genetics 1980; 94:115-133.
7. McGinnis W, Levine MS, Hafen E et al. A conserved DNA sequence in homeotic gene of the Drosophila Antennapedia and bithorax complexes. Nature 1984; 308:428-433.
8. Bürglin TR. Homeodomain proteins. In: Meyers RA, ed. Encyclopedia of Molecular Cell Biology and Molecular Medicine. Weinheim: Wiley-VCH Verlag GmbH and Co., 2005:179-222.
9. Gehring WJ, Affolter M, Bürglin T. Homeodomain proteins. Annu Rev Biochem 1994; 63:487-526.
10. Brooke NM, Garcia-Fernàndez J, Holland PWH. The ParaHox gene cluster is an evolutionary sister of the Hox gene cluster. Nature 1998; 392:920-922.
11. Kourakis MJ, Martindale MQ. Combined-method phylogenetic analysis of Hox and ParaHox genes of the Metazoa. J Exp Zoo (Mol Dev Evol) 2000; 288:175-191.
12. Ferrier DEK, Holland PWH. Ancient origin of the Hox gene cluster. Nature Rev Genet 2001; 2:33-38.
13. Finnerty JR, Martindale MQ. Ancient origins of axial patterning genes: Hox genes and ParaHox genes in the Cnidaria. Evol Dev 1999; 1(1):16-23.
14. Garcia-Fernàndez J. The genesis and evolution of homeobox gene clusters. Nature Rev Genet 2005; 6:1-12.
15. Collins AG. Evaluating multiple alternative hypotheses for the origin of Bilateria: An analysis of 18S rRNA molecular evidence. Proc Natl Acad Sci USA 1998; 95:15458-15463.
16. Peterson KJ, Eernisse DJ. Animal phylogeny and the ancestry of bilaterians: Inferences from morphology and 18S rDNA gene sequences. Evol Dev 2001; 3(3):170-205.
17. Schierwater B. My favorite animal, Trichoplax adhaerens. BioEssays 2005; 27:1294-1302.
18. Schierwater B, Kuhn K. Homology of Hox genes and the Zootype concept in early metazoan evolution. Mol Phyl Evol 1998; 9(3):375-381.
19. Gauchat D, Mazet F, Berney C et al. Evolution of Antp-class genes and differential expression of Hydra Hox/ParaHox genes in anterior patterning. Proc Natl Acad Sci USA 2000; 97:4493-4498.
20. Finnerty JR, Paulson D, Burton P et al. Early evolution of a homeobox gene: The ParaHox gene Gsx in the Cnidaria and Bilateria. Evol Dev 2003; 5(4):331-345.
21. Baguñà J, Riutort M. The dawn of bilaterian animals: The case of acoelomorph flatworms. BioEssays 2004; 26:1046-1057.
22. Garcia-Fernàndez J. Hox, ParaHox, ProtoHox: Facts and guesses. Heredity 2005; 94(2):145-152.

23. Pollard SL, Holland PWH. Evidence for 14 homeobox gene clusters in human genome ancestry. Curr Biol 2000; 10(17):1059-1062.
24. Minguillón C, Garcia-Fernàndez J. Genesis and evolution of the Evx and Mox genes and the extended Hox and ParaHox gene clusters. Genome Biol 2003; 4:R12.
25. Minguillón C, Gardenyes J, Serra E et al. No more than 14: The end of the amphioxusHox cluster. Int J Biol Sci 2005; 1:19-23.
26. Miller DJ, Miles A. Homeobox genes and the zootype. Nature 1993; 365:215-216.
27. Finnerty JR. Cnidarians reveal intermediate stages in the evolution of Hox clusters and axial complexity. Am Zoologist 2001; 41(3):608-620.
28. Cook CE, Jiménez E, Akam M et al. The Hox gene complement of acoel flatworms, a basal bilaterian clade. Evol Dev 2004; 6(3):154-163.
29. Aguinaldo AA, Turbeville JM, Linford LS et al. Evidence for a clade of nematodes, arthropods and other moulting animals. Nature 1997; 387:489-493.
30. Giribet G. Current advances in the phylogenetic reconstruction of metazoan evolution: A new paradigm for the Cambrian explosion? Mol Phyl Evol 2002; 24:345-357.
31. Telford MJ, Copley RR. Animal phylogeny: Fatal attraction. Curr Biol 2005; 15(8):R296-R299.
32. De Rosa R, Grenier JK, Andreeva T et al. Hox genes in brachiopods and priapulids and protostome evolution. Nature 1999; 399:772-776.
33. Balavoine G, de Rosa R, Adoutte A. Hox clusters and bilaterian phylogeny. Mol Phyl Evol 2002; 24:366-373.
34. Powers TP, Amemiya CT. Evidence for a Hox14 paralog group in vertebrates. Curr Biol 2004; 14(5):R183-R184.
35. Kobayashi M, Furuya H, Holland PWH. Dicyemids are higher animals. Nature 1999; 401:762-763.
36. Passamaneck YJ, Halanych KM. Evidence from Hox genes that bryozoans are lophotrochozoans. Evol Dev 2004; 6:275-281.
37. Telford MJ. Turning Hox "signatures" into synapomorphies. Evol Dev 2000; 2(6):360-364.
38. Telford MJ. Affinity for arrow worms. Nature 2004; 431:254-256.
39. Papillon D, Perez Y, Caubit X et al. Identification of chaetognaths as protostomes is supported by the analysis of their mitochondrial genome. Mol Biol Evol 2004; 21(11):2122-2129.
40. Helfenbein KG, Fourcade HM, Vanjani RG et al. The mitochondrial genome of Paraspadella gotoi is highly reduced and reveals that chaetognaths are a sister group to protostomes. Proc Natl Acad Sci USA 2004; 101:10639-10643.
41. Papillon D, Perez Y, Fasano L et al. Hox gene survey in the chaetognath Spadella cephaloptera: Evolutionary implications. Dev Genes Evol 2003; 213:142-148.
42. Van Auken K, Weaver DC, Edgar LG et al. Caenorhabditis elegans embryonic axial patterning requires two recently discovered posterior-group Hox genes. Proc Natl Acad Sci USA 2000; 97:4499-4503.
43. Halanych KM. Testing hypotheses of chaetognath origins: Long branches revealed by 18S ribosomal DNA. Syst Biol 1996; 45:223-246.
44. Ferrier DEK, Minguillón C, Holland PWH et al. The amphioxus Hox cluster: Deuterostome posterior flexibility and Hox14. Evol Dev 2000; 2(5):284-293.
45. Cameron RA, Rowen L, Nesbitt R et al. Unusual gene order and organization of the sea urchin Hox cluster. J Exp Zoo (Mol Dev Evol) 2005; 304B:1-14.
46. Martinez P, Rast JP, Arenas-Mena C et al. Organization of an echinoderm Hox gene cluster. Proc Natl Acad Sci USA 1999; 96:1469-1474.
47. Long S, Martinez P, Chen WC et al. Evolution of echinoderms may not have required modification of the ancestral deuterostome Hox gene cluster: First report of PG4 and PG5 Hox orthologues in echinoderms. Dev Genes Evol 2003; 213:573-576.
48. Mito T, Endo K. PCR survey of Hox genes in the crinoid and ophiuroid: Evidence for anterior conservation and posterior expansion in the echinoderm Hox gene cluster. Mol Phyl Evol 2000; 14(3):375-388.
49. Long S, Byrne M. Evolution of the echinoderm Hox gene cluster. Evol Dev 2001; 3(5):302-311.
50. Peterson KJ. Isolation of Hox and ParaHox genes in the hemichordate Ptychodera flava and the evolution of deuterostome Hox genes. Mol Phyl Evol 2004; 31:1208-1215.
51. Bourlat SJ, Nielsen C, Lockyer AE et al. Xenoturbella is a deuterostome that eats molluscs. Nature 2003; 424:925-928.
52. Duboule D. Temporal colinearity and the phylotypic progression: A basis for the stability of a vertebrate Bauplan and the evolution of morphologies through heterochrony. Development 1994; (Suppl):135-142.
53. Ferrier DEK, Holland PWH. Ciona intestinalis ParaHox genes: Evolution of Hox/ParaHox cluster integrity, developmental mode, and temporal colinearity. Mol Phyl Evol 2002; 24:412-417.

54. Ferrier DEK, Minguillón C. Evolution of the Hox/ParaHox gene clusters. Int J Dev Biol 2003; 47:605-611.
55. Crawford M. Hox genes as synchronized temporal regulators: Implications for morphological innovation. J Exp Zoo (Mol Dev Evol) 2003; 295B:1-11.
56. Patel NH. Time, space and genomes. Nature 2004; 431:28-29.
57. Seo HC, Edvardsen RB, Maeland AD et al. Hox cluster disintegration with persistent anteroposterior order of expression in Oikopleura dioica. Nature 2004; 431:67-71.
58. De Rosa R, Prud'homme B, Balavoine G. Caudal and even-skipped in the annelid Platynereis dumerilii and the ancestry of posterior growth. Evol Dev 2005; 7(6):574-587.
59. Pierce RJ, Wu W, Hirai H et al. Evidence for a dispersed Hox gene cluster in the platyhelminth parasite Schistosoma mansoni. Mol Biol Evol 2005; 22:2491-2503.
60. Arenas-Menas C, Martinez P, Cameron RA et al. Expression of the Hox gene complex in the indirect development of a sea urchin. Proc Natl Acad Sci USA 1998; 95:13062-13067.
61. Duboule D. Hox is in the hair: A break in colinearity? Genes Devel 1998; 12:1-4.
62. Roelen BAJ, de Graaff W, Forlani S et al. Hox cluster polarity in early transcriptional availability: A high order regulatory level of clustered Hox genes in the mouse. Mech Devel 2002; 119:81-90.
63. Suemori H, Noguchi S. HoxC cluster genes are dispensable for overall body plan of mouse embryonic development. Dev Biol 2000; 220:333-342.
64. Amores A, Suzuki T, Yan YL et al. Developmental roles of pufferfish Hox clusters and genome evolution in ray-finned fish. Genome Res 2004; 14:1-10.
65. Chambeyron S, Bickmore W. Chromatin decondensation and nuclear reorganization of the HoxB locus upon induction of transcription. Genes Devel 2004; 18:1119-1130.
66. Chambeyron S, Da Silva NR, Lawson KA et al. Nuclear reorganisation of the Hoxb complex during mouse embryonic development. Development 2005; 132:2215-2223.
67. Kmita M, van der Hoeven F, Zákány J et al. Mechanisms of Hox gene colinearity: Transposition of the anterior Hoxb1 gene into the posterior HoxD complex. Genes Devel 2000; 14:198-211.
68. Wada H, Garcia-Fernàndez J, Holland PWH. Colinear and segmental expression of amphioxus Hox genes. Dev Biol 1999; 213:131-141.
69. Duncan I. The Bithorax complex. Ann Rev Genet 1987; 21:285-319.
70. Averof M, Dawes R, Ferrier D. Diversification of arthropod Hox genes as a paradigm for the evolution of gene functions. Sem Cell Dev Biol 1996; 7:539-551.
71. Sharpe J, Nonchev S, Gould A et al. Selectivity, sharing and competitive interactions in the regulation of Hoxb genes. EMBO J 1998; 17(6):1788-1798.
72. Von Allmen G, Hogga I, Spierer A. Splits in fruitfly Hox gene complexes. Nature 1996; 380:116.
73. Negre B, Ranz JM, Casals F et al. A new split of the Hox gene complex in Drosophila: Relocation and evolution of the gene labial. Mol Biol Evol 2003; 20:2042-2054.
74. Lewis EB, Pfeiffer BD, Mathog DR et al. Evolution of the homeobox complex in the Diptera. Curr Biol 2003; 13(15):R587-R588.
75. Negre B, Casillas S, Suzanne M et al. Conservation of regulatory sequences and gene expression patterns in the disintegrating Drosophila Hox gene complex. Genome Research 2005; 15:692-700.
76. Powell JR, DeSalle R. Drosophila molecular phylogenies and their uses. Evol Biol 1995; 28:87-139.
77. Yasukochi Y, Ashakumary LA, Wu C et al. Organization of the Hox gene cluster of the silkworm, Bombyx mori: A split of the Hox cluster in a nonDrosophila insect. Dev Genes Evol 2004; 214:606-614.
78. Orlando V. Polycomb, epigenomes, and control of cell identity. Cell 2003; 112:599-606.
79. Kmita M, Kondo T, Duboule D. Targeted inversion of a polar silencer within the HoxD complex reallocates domains of enhancer sharing. Nature Genetics 2000; 26:451-454.
80. Belozerov VE, Majumder P, Shen P et al. A novel boundary element may facilitate independent gene regulation in the Antennapedia complex of Drosophila. EMBO J 2003; 22(12):3113-3121.
81. Ronshaugen M, Biemar F, Piel J et al. The Drosophila microRNA iab-4 causes a dominant homeotic transformation of halteres to wings. Genes Dev 2005; 19:2947-2952.
82. Stark A, Brennecke J, Bushati N et al. Animal microRNAs confer robustness to gene expression and have a significant impact on 3'UTR evolution. Cell 2005; 123:1133-1146.
83. Falciani F, Hausdorf B, Schröder R et al. Class 3 Hox genes in insects and the origin of zen. Proc Natl Acad Sci USA 1996; 93:8479-8484.
84. Stauber M, Jäckle H, Schmidt-Ott U. The anterior determinant bicoid of Drosophila is a derived Hox class 3 gene. Proc Natl Acad Sci USA 1999; 96:3786-3789.
85. Cameron CB, Garey JR, Swalla BJ. Evolution of the chordate body plan: New insights from phylogenetic anlyses of deuterostome phyla. Proc Natl Acad Sci USA 2000; 97:4469-4474.
86. Jakob W, Sagasser S, Dellaporta S et al. The Trox-2 Hox/ParaHox gene of Trichoplax (Placozoa) marks an epithelial boundary. Dev Genes Evol 2004; 214:170-175.

Chapter 5

Evolutionary Genomics of *Hox* Gene Clusters

Sonja J. Prohaska, Peter F. Stadler and Günter P. Wagner*

Abstract

The evolution of *Hox* clusters in vertebrates follows different patterns than those of in-vertebrate clusters. More stringent structural constraints in vertebrates are apparent from tighter cluster organization and the systematic expulsion of repetitive material. We speculate that the tendency of vertebrates to maintain Hox clusters after genome duplications might be related to these stricter constraints. Duplications may temporarily lift these constraints, thereby opening a window of adaptive opportunity for functional differentiation of the *Hox* genes that eventually leads to their fixation.

Introduction

Hox genes were the first family of developmental genes who were recognized as being shared among widely different animals.[1] This made them the paradigmatic example for a new approach to evolutionary biology which promises to transform the way evolutionary biology can be pursued. The importance of this discovery has two dimensions, a technical and a conceptual. The technical dimension is that homologs of *Drosophila Hox* genes where found in mouse because of a conserved sequence motif, called the homeobox. The conservation of these sequence motives among very distantly related animals offers the opportunity to clone and study developmentally relevant genes in a large number of nonmodel organisms. Hence it is possible to study development not only in a hand full of phylogenetically ill-placed model organisms, arguably unsuitable for rigorous evolutionary inferences. Instead, we now can study the developmental genes of almost any species we chose to compare.

Of equal importance as the technical advance is the conceptual continuity established among animals with different body plans through the discovery of a shared developmental-genetic machinery. Ever since biological diversity became the subject of scientific study it was clear that extant diversity comes in two kinds: at the one hand the rich gradual diversity with a clear hierarchical structure of species sharing the same basic structure or body plan, as the insects and the fishes. On the other hand, there are life-forms which do not seem to have much, if anything, in common, like echinoderms and mollusks and vertebrates. Georges Cuvier[2] put them into separate "embranchments" basically recognizing the fact that the body structure is incomparable with anatomical means. Traditionally these life forms have been called "body plans", and animals with different body plans are classified in different phyla. The problem with understanding the evolution of body plans is that one needs very detailed knowledge of cytology, microanatomy and embryology to find any commonalities among these animals. For instance, the embryological origin of the mouth opening can be used to classify phyla into proto- and deuterostomes; The micro-structure of the flagellum and the adjacent cell cortex

*Corresponding Author: Günter P. Wagner—Department of Ecology and Evolutionary Biology, Center for Computational Ecology, Yale University, New Haven, Connecticut 06520, U.S.A. Email: gunter.wagner@yale.edu

HOX Gene Expression, edited by Spyros Papageorgiou. ©2007 Landes Bioscience and Springer Science+Business Media.

connects the single-celled choanoflagellates and multicellular animals. This situation changed radically by the realization that anatomically as fundamentally different animals as mice and flies still have recognizably homologous developmental genes and that their body axis is patterned by similar mechanisms. Not only did this confirm continuity among flies and mice in the sense of phylogenetic history but also holds the promise to lead to a mechanistic understanding of macro-evolutionary diversity. After all these genes are directly involved in the development of "body plan characters" and thus their evolution may be also linked to the evolution of body plan characters. Needless to say that we have not yet arrived at this level of understanding, but it is clear that it is critical to gain a deeper understanding of the evolution of these genes and the forces that shape them to eventually arrive at a narrative that connects even the most dissimilar organisms into a unified picture of nature.

Our objective in this chapter is to summarize the current knowledge of the molecular evolution and genomics of *Hox* gene clusters with a special emphasis on the evolution of *Hox* clusters in chordates. For a review of the early history of *Hox* gene clusters, see the chapter by Ferrier in this volume.

Hox gene clusters have been a particularly charismatic example for the evolution of developmental genes and their connection to body plan evolution. The differences between the *Drosophila* and the mouse *Hox* gene families are suggestive of evolutionary progress. In *Drosophila* the *Hox* genes are found in two closely linked clusters on the same chromosome, which is a situation that was later found to be derived form a single cluster.[3] In the mouse, in contrast, it was found that there are 39 *Hox* genes in four clusters on four different chromosomes.[4] This difference in *Hox* gene number and organization makes sense in the light of the perceived higher complexity of the mammalian body plan as compared to that of a fly. This view was further reinforced by reports that suggested a *scala naturae* like relationship between the number of *Hox* clusters and body plan complexity with one cluster in nonchordates, two in Amphioxus, three in lamprey and four in mammals and other gnathostomes.[5] This and similar reports, however, were based on PCR surveys or Homeobox fragments.[6] These methods are very useful as a first view of the *Hox* gene complement of a new species but are inherently inaccurate when it comes to the estimation of total *Hox* gene and *Hox* cluster number. It soon became clear that the situation is much more complex and that a simple linear relationship between perceived body plan complexity and *Hox* cluster number does not describe the situation. Amphioxus was found to have only a single *Hox* cluster, although with 14 genes,[7] zebrafish and fugu were found to have seven *Hox* clusters.[8,9] The situation in lamprey is still unresolved, in spite of major efforts by leading labs in the field (see below), and the correlation between major evolutionary events in vertebrate phylogeny and *Hox* cluster evolution is questionable,[10] although in the ray-finned fishes a close association between *Hox* cluster duplication and the teleost radiation has been established.[11]

When one studies the evolution of *Hox* genes it is useful to keep three basic insights in mind:

1. *Hox* cluster and *Hox* gene number increase is not necessary for evolutionary innovations.

This was first shown by Sean Carroll[12] who pointed out that insects and other invertebrates are very productive in terms of body plan innovations and evolutionary success without an obvious connection to *Hox* cluster duplications. Even in vertebrates the arguably most momentous evolutionary advances, like the origin of terrestrial forms (tetrapods), mammals and birds occurred without an expansion of the *Hox* gene complement. That does not mean that *Hox* genes played no role in these events. In the case of mammals they certainly experienced directional selection (see e.g., ref. 13), but it is not necessarily the number of *Hox* genes and clusters that mattered in these events.

2. *Hox* gene duplication in itself does not cause evolutionary innovations.

For instance, the genomes have been duplicated multiple times in the genus Xenopus,[14] an additional duplication happened in the salmonids[15] but in neither case is there an association with an evolutionary novelty.

3. *Hox* clusters are often duplicated, but only a small fraction of the duplicated clusters has been maintained.

In animal phylogeny *Hox* genes and *Hox* clusters duplicate at least as often as genomes duplicate, but only a small fraction of these events led to the permanent establishment of additional *Hox* clusters. An-amniotes have a very high frequency of genome duplications, particularly well documented in fish and amphibians, but so far only up to 8 clusters created in three different duplication events have been found. There are other *Hox* clusters found in fishes, as in Salmonids[16,17] and paddlefish, a basal ray-finned fish (Crow, Amemiya and Wagner in preparation), but these are relatively recent events and it is not clear whether they will be retained in the long run.

Clearly the evolutionary history of *Hox* genes in vertebrates and their role in body plan evolution are far from understood, and maybe we even need to rethink the questions we are asking about their evolution.[18,19]

Genomic Features

As pointed out above, it is clear that *Hox* gene genomics has no simple relationship to developmental evolution. It is necessary to understand the history and dynamics of *Hox* cluster evolution in its own right before we can sort out its contribution to other evolutionary phenomena like the evolution of body plans. Here we review differences in the variational tendencies and genomic organization between chordate *Hox* clusters and that of most other invertebrate *Hox* clusters, which may account for the different evolutionary dynamics found in chordates as compared to most of the remaining evolutionary history.

Loss of Hox-*Cluster Integrity in Invertebrates*

There is a striking difference between the tightly linked compact *Hox* clusters of all vertebrates studied so far and the much looser organization of this gene system in invertebrate species. Indeed, while all gnathostome *Hox* clusters appear to be intact, an unfragmented and uninterrupted lineup of *Hox* genes as in the cephalochordate *Branchiostoma floridae*[20] seems to be the exception rather than the rule in invertebrates.

In some lineages we observe a massive loss of *Hox* genes. The nematode *Caenorhabditis elegans*, for example, has only 6 *Hox* genes arranged in 3 separated pairs that are spread widely across 5Mb of chromosome III. The gene loss is restricted to the nematode lineage, as nematomorphs have 11 genes, including the full set of at least 9 protostome *Hox* genes.[21,22]

At least three different splits have been reported in the *Hox* clusters of various Drosophilids, see refs. 23,24 and the references therein: Antp//Ubx, Ubx//Abd-A, and lab//pb. In these cases, however, the loss of cluster integrity is not associated with a loss of *Hox* genes. A similar situation has been observed in *Bombyx mori*.[25]

As a representative of the lophotrochozoans, only the parasite *Schistosoma mansoni* has been studied w.r.t. cluster integrity.[26] Using FISH it was show that three *Hox* genes (*SmHox1*, *SmHox4*, *SmHox8*) localize to chromosome 3 and the long arm of chromosome 4. In silico analysis of the available genomic data shows a presumably unduplicated schistosome *Hox* cluster both dispersed and disintegrated in the genome.

In contrast to the canonical *Hox* cluster of the cephalochordate *Branchiostoma floridae*, all other basal deuterostome lineages studied so far show drastic rearrangements. In the echinoderm *Strongylocentrotus purpuratus* there is a tightly linked cluster which, however, has undergone a series of internal rearrangements (Fig. 1).[27]

In urochordates, the cluster has disintegrated with fragments distributed over two chromosomes in *Ciona intestinalis*.[28-30] An extreme case is the larvacean *Oikopleura dioica*, which shows a complete loss of *Hox* cluster integrity.[31] In this species all *Hox* genes are separated at least several 100kb from each other. Furthermore, there is a lineage specific duplication of *Hox9* and several other homeodomain proteins which could reflect a rediversification of the homeobox gene complement following major group losses.[32]

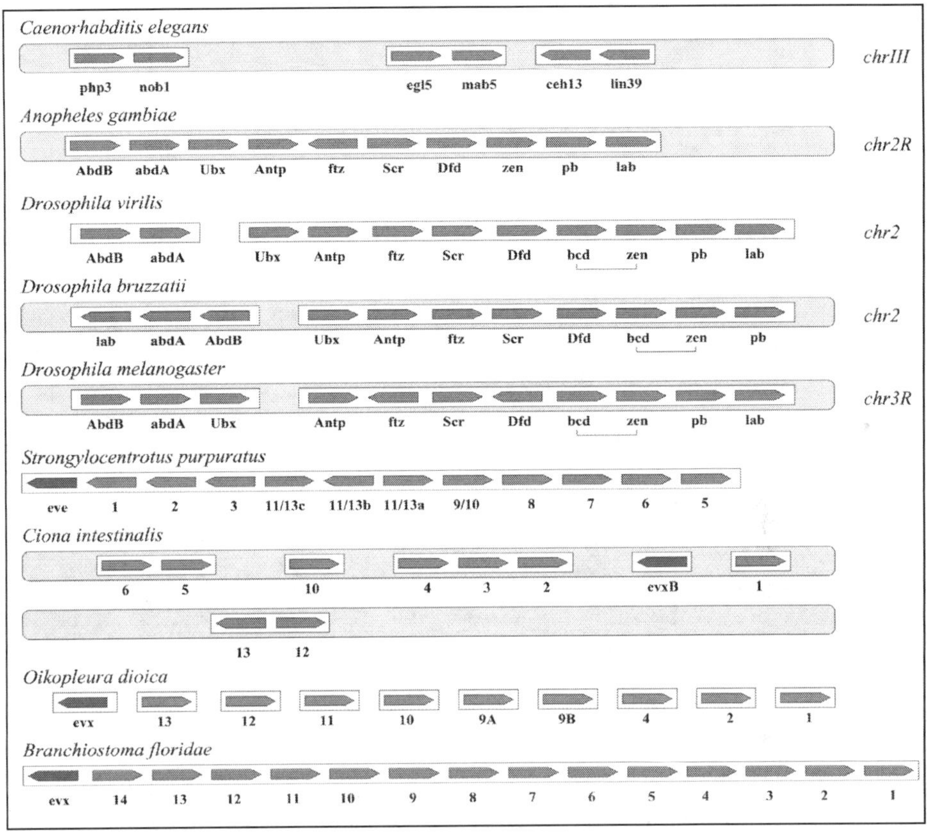

Figure 1. Examples of *Hox* gene systems that lost cluster integrity. Schematic drawings are not to scale. Grey bar indicate chromosomes where known, white boxes indicate tightly linked sections. For unfinished genomes, white boxes imply location of the genes on the same scaffold in the assemblies. The direction of the individual scaffolds on the *Ciona intestinalis* chromosomes is arbitrary, the orientation within each scaffold is correct. In *Oikopleura dioica*, all *Hox* genes are located on different scaffolds.

Repetitive Elements: Differences between Vertebrates and Insects

In contrast to their invertebrate counterparts, gnathostome *Hox* gene are compact (100kb instead of 400kb in Amphioxus or almost 1Mb in the sea urchin), uninterrupted, and even exhibit highly conserved intergenic distances, (see e.g., refs. 33-35). These facts suggest that the gnathostome *Hox* clusters have to satisfy much tighter organizational constraints than their invertebrate counterparts. Strikingly, the vertebrate *Hox* loci are part of highly conserved extended syntenic blocks.[36]

The distribution of interspersed repetitive elements at the genomic locus of the *Hox* clusters reflects its organizational constraints. Repetitive elements are dramatically depleted in the core of *Hox* clusters compared to the regions upstream and downstream of the *Hox* cluster.[37] Figure 2 displays the situation for the *HoxA* cluster of the armadillo as an example.

Two characteristics of mobile genetic elements might be responsible for this effect. First, repetitive elements are a base for chromosomal rearrangements. An enhanced frequency of transposon-mediated inversions in *Drosophila*[38] was proposed as a possible cause for the fragmentation of the *Drosophila Hox* cluster.[39] In addition, transposable elements have been

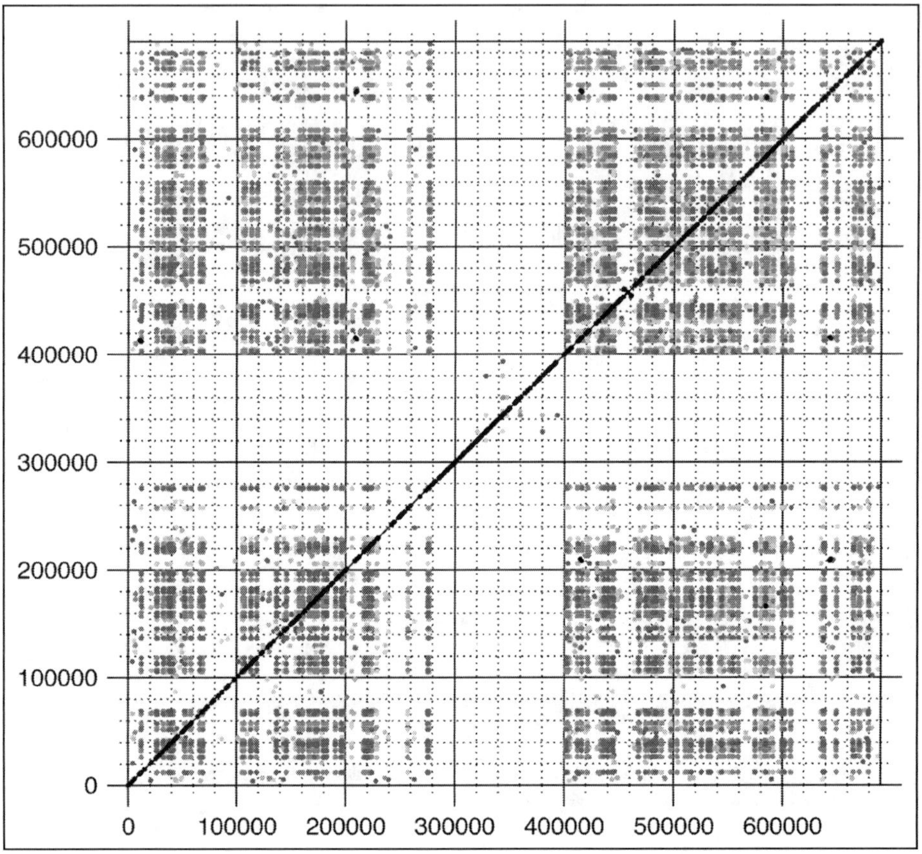

Figure 2. Dot plot of the armadillo ENDCODE region EN010 including the *HoxA* cluster. The cluster is located at the region from 249000 (*evx1*), 293000 (*hoxA13*) to 394000 (*hoxA1*). The sequence similarity is denoted by the color from high similarity (black, violett, red) to low similarity (green, blue, cyan). Off-diagonal dots indicate repetitive elements. The off-diagonal dots within the *Hox* cluster show the similarity of the homeodomains. A color version of this figure is available online at www.Eurekah.com.

reported close to *Hox* genes in several of the organisms with fragmented *Hox* clusters discussed above. Second, repetitive elements can show regulatory activity and this might influence *Hox* gene expression. For instance, Human *Alu* elements often function as RNA polymerase III promoters. In some cases the regulatory abilities of mobile DNA elements are used by the host.[40,41] In general, however, we expect that any interference with the tight cross-regulatory network of a *Hox* cluster will be detrimental to its function.

Further evidence for negative selective pressure to exclusion of repetitive sequences comes from the length of repetitive elements in *Hox* clusters. In the clusters with highly reduced repeat density, the remaining repeats are also shorter independent of the type of the repetitive elements.[37]

A more detailed analysis shows that repetitive DNA is not evenly distributed in gnathostome *Hox* clusters. They predominately accumulate in regions where *Hox* genes have been lost in the aftermath of the 2R genome duplication, namely in the IGR between *hoxB13* and *hoxB9*, at the 3' ends of the *HoxC* and *HoxD* clusters, as well as at the 5' end of the of the *HoxA* cluster, (Fig. 3). The density of repeats in the intergenic regions between *hoxB13* and *hoxB9* is almost

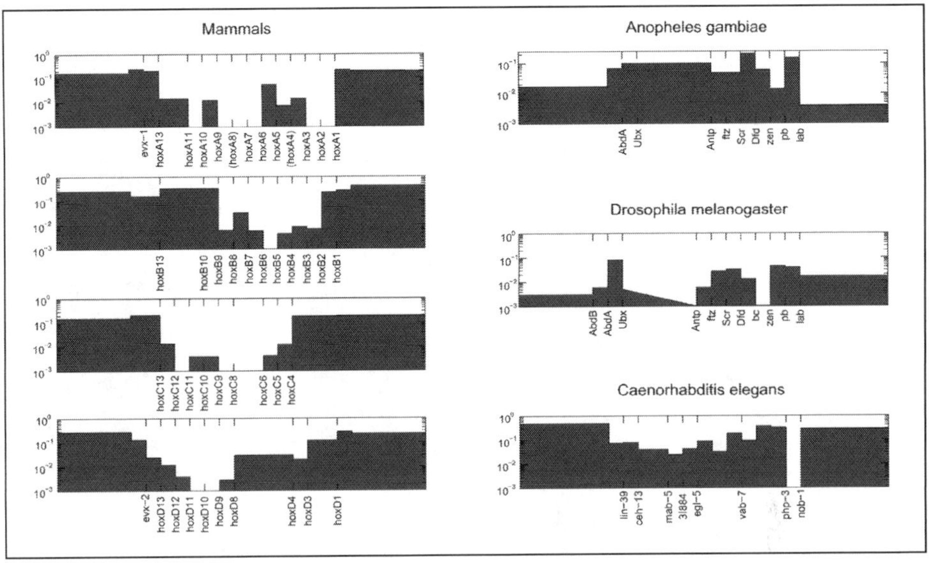

Figure 3. Distribution of the fraction repetitive elements in the intergenic regions of gnathostome *Hox* clusters. The density of repeats averaged over 100000nt up- and downstream of the cluster and for the intergenic regions between the indicated *Hox* genes. The "invasion" of repeats from the cluster ends is clearly visible in the mammalian clusters (left). The protostome clusters (right) do not show repeat exclusion. The few repeat-free intergenic regions in the protostome clusters are probably counting artifacts since the corresponding sequence intervals are very short: 286nt between *php-3* and *nob-1* in *C. elegans,* 2531nt between *bc* and *zen* in *D. melanogaster.* (Updated from ref. 37.)

the same as in the regions adjacent to the cluster. The same effect can be observed for the recently duplicated *Hox* clusters in teleost fishes (Fig. 12).

Tight cluster organization as highlighted by the exclusion of repetitive sequence elements may in fact be a gnathostome innovation since a significant reduction of repetitive sequences can be observed only in gnathostome lineages.

Noncoding RNA within the Hox *Clusters*

It has become clear over the last year or two that only a tiny fraction of the transcriptional output of the human genome are protein-coding mRNAs.[42-47] Most of these novel transcript have limited or no protein-coding potential.

Antisense transcription has been implicated in various mechanism of gene regulation, including RNAi-like degradation of the corresponding sense transcripts, gene silencing at the chromatin level, and by means of competition of sense and antisense transcription. Global transcriptome analysis[52] shows that a large proportion of the genome can indeed produce transcripts from both strands. A recent map of the human *HoxA* cluster based on data from the ENCODE project shows that the number of anti-sense transcripts in this region rivals the number of sense genes (Fig. 4). EST data, where available, suggest a similar picture for the other mammalian *Hox* clusters. The evolution of these non*Hox* members of the *Hox* clusters has not been studied systematically so far.

The mammalian *HoxA* cluster contains one of the better-known antisense transcripts, the *hoxA11-AS* gene, which is located upstream of *hoxA11*. This gene is conserved among mammals[53] and appears to functions by transcriptional interference, repressing HOXA11 expression by competing for transcription, rather than by sense/anti sense interaction.[54] Just as the

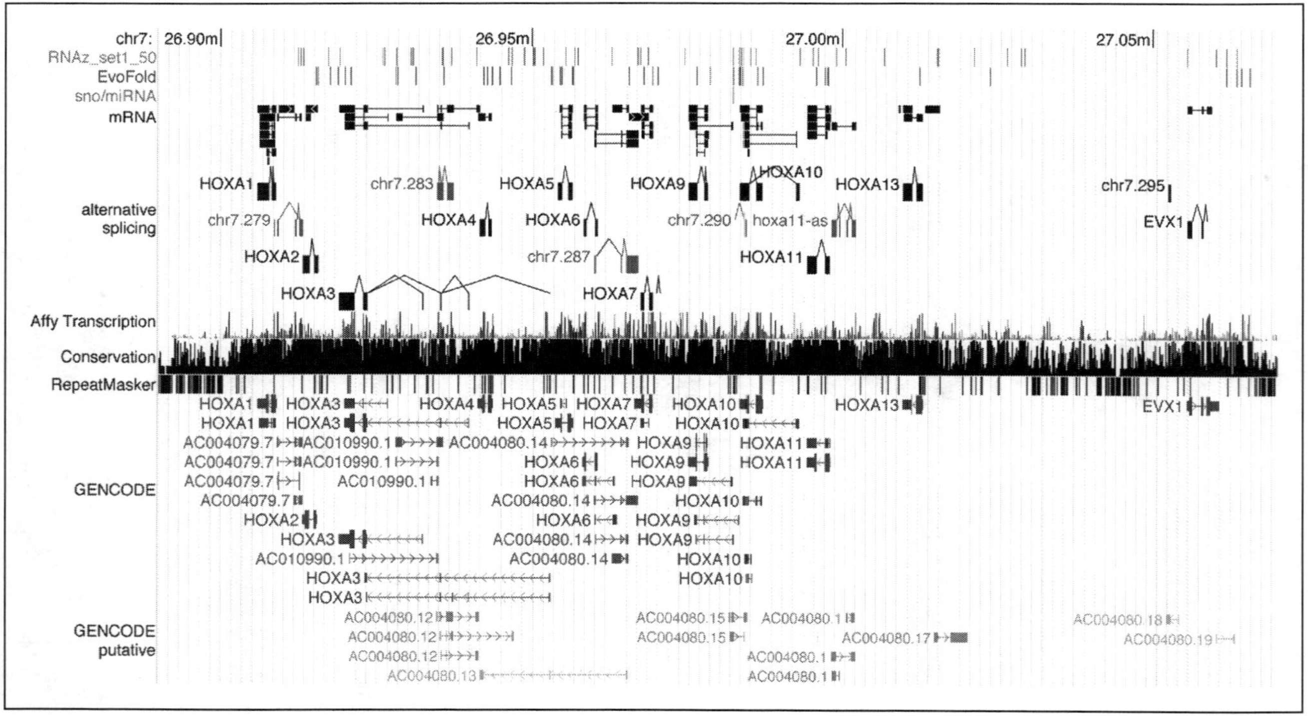

Figure 4. Map of the human *HoxA* cluster extracted from the UCSC Genome Browser (Feb 2006) based on hg17 coordinates. The top two lines show predicted structured RNA by RNAz[48] outside coding sequences and by evofold.[49] The letter method is less sensitive but the track also includes coding regions. Below mRNA structures from GENBANK[50] and the alternative splicing track based on ref. 51 gives an overview of transcript structure. Genes on the opposite strand relative to the *Hox* genes are shown in a reddish color. The blue track summarizes Affymetrix transfrag data.[47] Below conservation and repeat track GENECODE annotation based on Sanger's HAVANA annotation method are summarized. Well-supported, experimentally verified (dark) and putative genes (light colors) are shown separately. With the exception of a single splice variant of *AC004079.7*, no non*Hox* gene in this region has coding potential. A color version of this figure is available online at www.Eurekah.com.

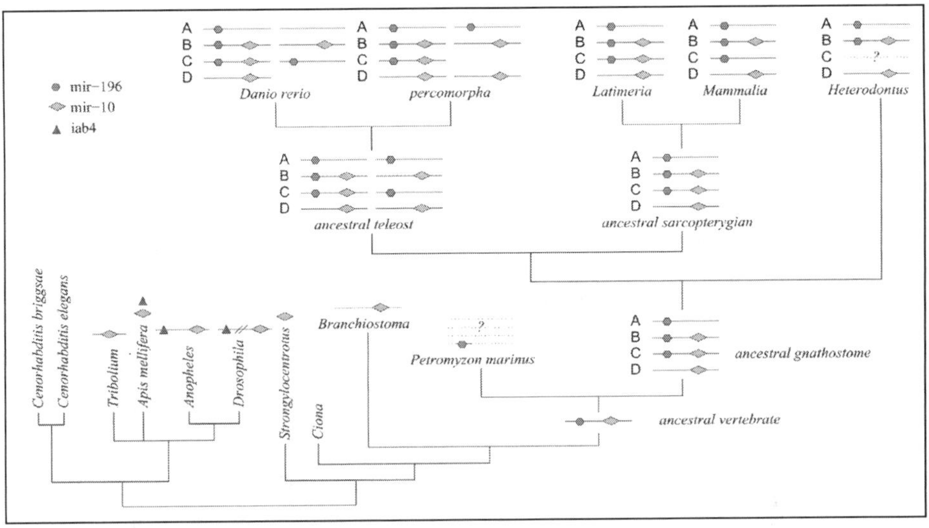

Figure 5. Evolution of microRNAs located within the *Hox* cluster (updated from ref. 57). Medaka appears to have lost the *mir-196-Ba* gene which is present in Danio, Fugu, and Tetraodon.

Hox genes themselves, the anti-sense transcripts in general are processed in several splice-variants. A prominent example in the *HoxB* locus is the small (6 kDa) nuclear protein PRAC, which is located about 4kb *down*stream of *hoxB13* on the *same* strand. The PRAC gene encodes a 382 RNA found in prostate, rectum, distal colon, and in three prostate cancer cell lines[55] and is at present the only known protein coding gene to have "invaded" a mammalian *Hox* cluster in sense direction. A related gene, *PRAC2*, is located between *PRAC* and *hoxB13* on the opposite strand.[56]

Few "classical" RNAs are located within the *Hox* clusters: Three families of microRNAs are known to be located within the *Hox* gene clusters (see Fig. 5).[57,58] The known *mir-10* sequences are located between *hox5* and *hox4* in vertebrates and, correspondingly, between *Dfd* and *Scr* in arthropods. Of all genomes considered here, *mir-10* was absent only from the two nematodes and from urochordates.

Vertebrate homologs of *mir-10* were found in *HoxB*, *HoxC*, and *HoxD* clusters of gnathostomes, while they are absent in all investigated *HoxA* clusters including shark, latimeria, bichir, various teleosts and tetrapods. The *mir-10* copy in the *HoxC* is present only in teleosts, Xenopus and Latimeria. A *mir-10* homolog was identified in the sea urchin *Strongylocentrotus purpuratus*. As expected it is found upstream of *hox3*[27] lending credibility to the idea that *mir-10* is associated with an evolutionarily conserved *hox3*-transcript that has its transcription start *upstream* of *hox4* (which is lost in sea urchins).[59,57]

The second family of microRNAs, *mir-196*, is located upstream of *hox9* and regulates *hox8* mRNAs.[60] No invertebrate homologs of *mir-196* were found, the oldest representative of this family is a lamprey *mir-196* sequence located downstream of the PG10 gene *HoxW10a* in *Petromyzon marinus*.[61] All gnathostome *mir-196* sequences are located in the *HoxA*, *HoxB*, and *HoxC* clusters, while no candidates were detectable in any of the available *HoxD* cluster sequences.

In insects, the unrelated microRNA *iab-4* is analogous to *mir-196*.[60] It is located in the corresponding region between *AbdB* and *AbdA* and interacts with *Ubx*.[62] We find that this sequence is conserved in a larger group of insect species but probably not even throughout the arthropod clade.

Chordate *Hox* Cluster History: How Little Do We Know?

Is the Amphioxus Cluster Ancestral?

See contribution by D.E.K. Ferrier in this book for a discussion of the evolution of *Hox* clusters in basal deuterostomes, in particular echinoderms and hemichordates.

The closest relatives of vertebrates are the cephalochordates, i.e., Amphioxus and its relatives, and the hemichordates. Traditionally the cephalochordates have been considered the sister taxon of vertebrates forming the chordate clade. This view has been challenged recently claiming that the hemichordates are closer to vertebrates than the cephalochordates.[63] For the discussion of the ancestral *Hox* clusters of vertebrates this ambiguity is irrelevant since the hemichordate *Hox* clusters are so derived (see section "What Happened in Agnate Vertebrates?") that they give no information about the ancestral situation for vertebrates. The only *Hox* cluster organization that could be representative of the ancestral situation is that of Amphioxus.

Amphioxus has been found to have a single 400kb cluster with 14 *Hox* genes. It is pretty clear that the *Hox* genes 1 to 10 are orthologs of the corresponding paralog groups in the gnathostomes, but the situation is less clear with the Abd-B related genes, PG 11, 12 and 13. The traditional criterion for orthology based on gene tree reconstruction does not support the hypothesis that the Amphioxus genes 11, 12 and 13 are orthologous to the PG 11 to 13. In the gene tree analysis the Amphioxus and the vertebrate genes form separate clades (Fig. 6) suggesting that the 5' genes originated independently in the cephalochordate and in the gnathostome stem lineage. However, the recent discovery of *hox14* paralog group genes in basal gnathostome lineages changes the balance of evidence in favor of the hypothesis that the Amphioxus *Hox* cluster represents the organization of the ancestral vertebrate *Hox* cluster. PG 14 genes are different from other chordate *Hox* genes as they have an intron in the Homeobox. This is presumably the reason why this group was not discovered earlier, since this intron affects the target sequence of the "universal" Homeobox primers[65,66] and thus prevents the detection of these genes in PCR surveys. Powers and Amemiya[67] have shown that there is a *Hox* gene 5' of the *hoxA13* in Latimeria and of *hoxD13* in shark which is similar to the Amphioxus *hox14* gene. These PG14 genes have presumably been lost in the more derived gnathostomes. Their presence in basal lineages, however, makes it very likely that the Amphioxus *Hox* gene cluster structure and gene complement is representative of the ancestral vertebrate *Hox* cluster.

If the Amphioxus *Hox* genes are orthologs of the vertebrate paralog groups the question arises why the sequence comparison of the 5' *Hox* genes leads to the wrong gene tree. Peter Holland and his colleagues argue that the higher rate of 5' *Hox* gene evolution, a.k.a. terminal flexibility, leads to long branch attraction and thus to misleading phylogenetic signal.[7] An alternative has been advanced by Campos and colleagues[64] suggesting that the coevolution of *Hox* genes with their protein-protein interaction partners could also lead to the same reconstruction artifact. A better understanding of the protein-protein interaction partners of 5' *Hox* genes and their evolution will be necessary to distinguish between these alternatives.

What Happened in Agnate Vertebrates?

The earliest vertebrates found in the fossil record are jaw-less.[68] Only two jaw-less groups are still alive, the lamprey (*Hyperoartia*) and hagfish (*Hyperotreti*).[69] These lineages are considered recent offshoots of the large radiation of jaw-less vertebrates at the base of the vertebrate phylogeny.[70] Their phylogenetic affiliation, however, is unclear, both in terms of their relationship to the jaw-less groups of fossil vertebrates as well as their relationship to each other. On morphological grounds, lamprey and hagfish are often considered paraphyletic to jawed vertebrates.[71] In contrast, most molecular phylogenies suggest that lampreys and hagfish form a monophyletic clade which is the sister taxon to the jawed vertebrates,[70,72-77] with few dissenting voices.[78,79]

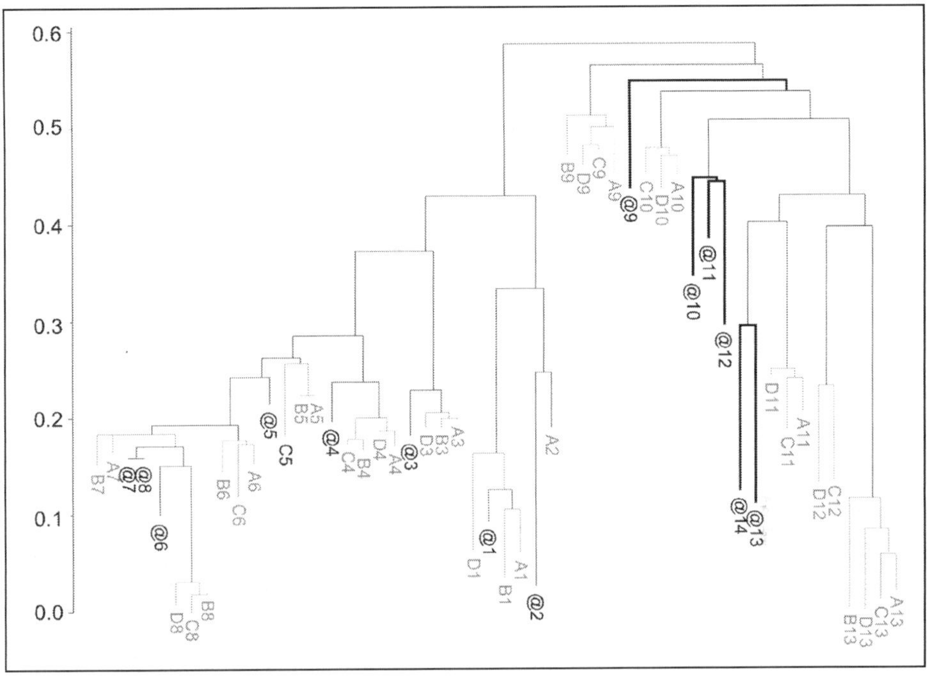

Figure 6. Neighbor-joining tree of the homeobox sequences of the *Hox* class transcriptions factors from Amphioxus (*Branchiostoma floridae,* marked by @) and for *Homo sapiens.* Gnathostomes clusters are labeled A through D. (Updated from ref. 64.)

PCR surveys[5,80] and genomic mapping data[61,81] of *Petromyzon marinus* as well as individual genes from other species (*Lethenteron japonicum*)[83] and (*Lampetra fluviatilis*)[84] indicate that lampreys have at least three, and probably four or more, *Hox* clusters (see Fig. 7 for details).

Despite recent efforts, the evolutionary history of the lamprey *Hox* genes and their relationship with the quadruplicate mammalian *Hox* clusters is far from being resolved, however. Together with evidence from other gene families[85-87] one can rather safely conclude, however, that lampreys *Hox* clusters have undergone at least one independent round of duplications. At present, the only firm conclusion is that lamprey have at least three *Hox* clusters. The difficulties in obtaining linkage information reported in refs. 61 and 81 suggest that *Hox* clusters might have broken up in lampreys, or, alternatively, that the known lamprey *Hox* genes stem from more than 3 or 4 clusters.

A PCR survey of the hagfish *Eptatretus stouti* provides evidence for at least 33 different *Hox* genes in the hagfish genome.[88] The number of homeobox fragments identified per paralog group is very variable and ranges from zero in group 12 to seven in group 9 (Fig. 8). This data is most consistent with the hypothesis of multiple clusters, but in itself can not provide evidence for physical linkage between the genes.

The small homeobox fragments from PCR surveys provide insufficient phylogenetic signal to reconstruct a detailed gene phylogeny. Nevertheless, at least two gene pairs are very likely first order paralogs[88] supporting the conclusion that independent *Hox* cluster duplications must have taken place in the hagfish lineage. This situation is similar to the findings about lamprey *Hox* genes, which, within a paralog group, are predominantly[5,61,80,81] or perhaps even exclusively derived within the lamprey lineage.[82] Hence it is possible that no *Hox* cluster duplication may have happened in the stem lineage of vertebrates, i.e., prior to the most recent common ancestor of Recent jaw-less and jawed vertebrates.

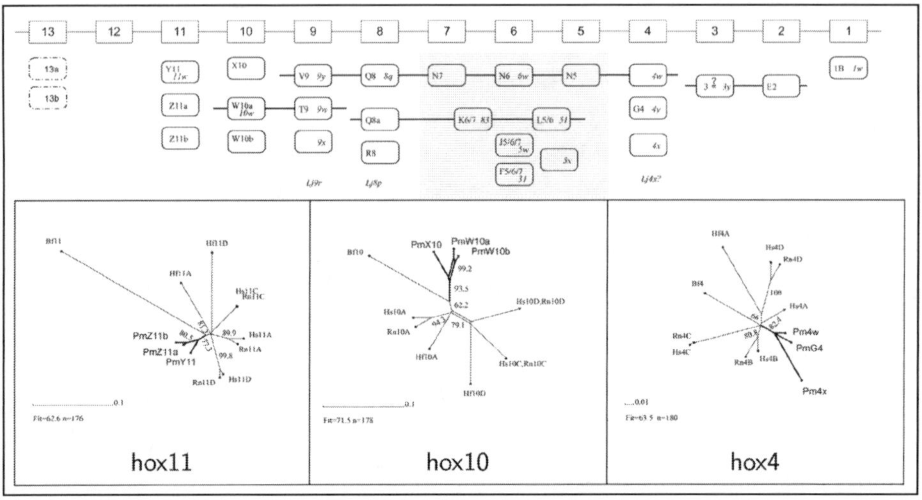

Figure 7. *Hox* clusters in *Petromyzon marinus*. Top) Summary of the known genes from PCR and cDNA data.[5,61,80-82] Dashed boxed indicate PCR data only, physical linkage is indicated by a line. The sequences of paralog groups 5, 6, and 7 are insufficient to resolve their mutual relationships. Recent studies on the origins of jaws were performed on different species (*Lethenteron japonicum*[83] and *Lampetra fluviatilis*).[84] The Lethenteron data possibly contain 2 or 3 additional paralogs. Below) Buneman graphs of the homeobox sequences for paralog groups 11, 10, and 4. The "fit" value is the fraction of total distance that is resolved into splits, the "split-prime" remainder can be interpreted as noise. Distances are given as relative frequency of mutations. Abbreviations: Hs: *Homo sapiens*, Hf: *Hetrodontus francisci* (horn shark), Bf: *Branchiostoma floridae* (Amphioxus), Pm: *Petromyzon marinus* (lamprey).

A Quartet Mapping analysis of the support for the three possible gene trees among Amphioxus, lamprey, hagfish and gnathostome *Hox* genes (Fig. 8), however, yields a surprising result: There is weak but distinctive support for the unrooted tree which associates some of the hagfish genes with gnathostome *Hox* genes: ((H,G)(A,L)). In contrast, there are only two rooted phylogenies for the major recent chordate taxa currently under discussion: (O(A(H(L,G)))), and (O(A((H,L)G))), which molecular phylogenies favoring a monophyletic status of a hagfish/lamprey clade. Either phylogenetic hypothesis together with the Quartet Mapping data and the results for lamprey summarized above have two implications. First they suggest that there was more than one *Hox* cluster in the most recent common ancestor of gnathostomes and cyclostomes. This conclusion is inferred from the significant association between hagfish *Hox* genes with gnathostome *Hox* genes. This interpretation is consistent with the 2R hypothesis.[90] Second, the results imply that the *Hox* gene situation in lamprey must be highly derived with many *Hox* gene lineages extinct and replaced with more recent duplicates.

A plausible scenario for the evolution of the *Hox* gene clusters runs as follows (Fig. 9):

1. The first duplication of the cephalochordate *Hox* cluster occurred in a common ancestor of gnathostomes, lampreys and hagfishes. The redundancy introduced in the duplication was resolved differently in these three groups.
2. While hagfishes and gnathostomes retained a large number of first order paralogs, the redundant genes were essentially lost in the lamprey lineage.
3. A further round of duplication lead to the four gnathostome clusters.
4. Two independent rounds of duplications occurred in both the lamprey and hagfish lineages.

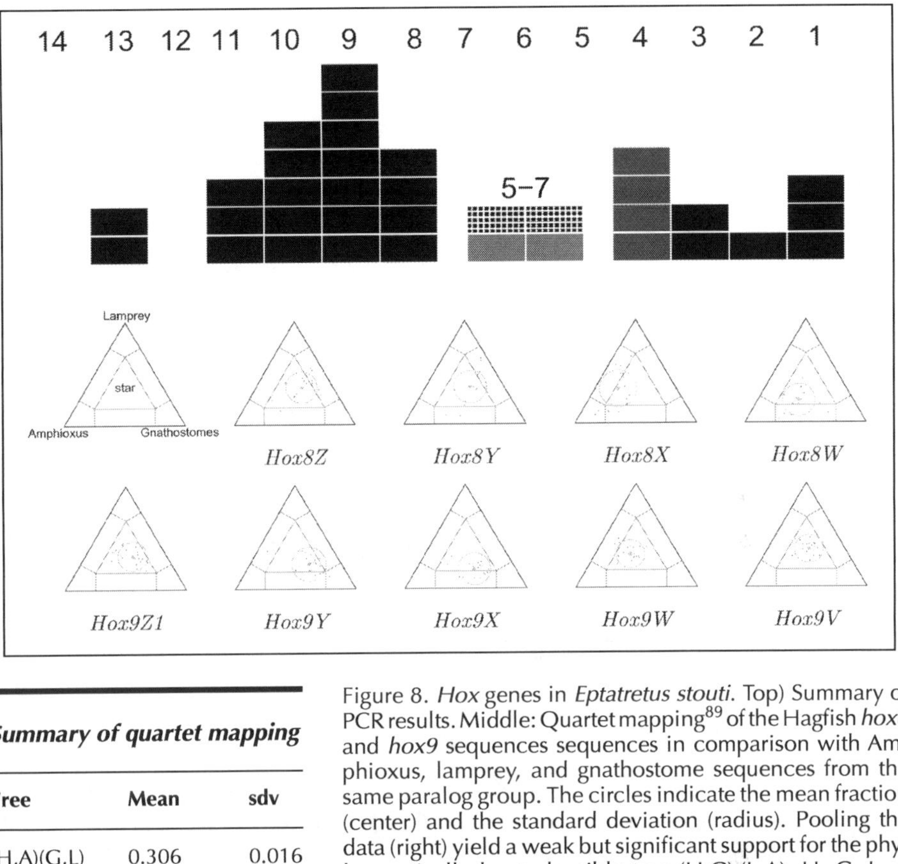

Summary of quartet mapping

Tree	Mean	sdv
(H,A)(G,L)	0.306	0.016
(H,G)(A,L)	0.367	0.013
(H,L)(G,A)	0.327	0.012

Figure 8. *Hox* genes in *Eptatretus stouti*. Top) Summary of PCR results. Middle: Quartet mapping[89] of the Hagfish *hox8* and *hox9* sequences sequences in comparison with Amphioxus, lamprey, and gnathostome sequences from the same paralog group. The circles indicate the mean fraction (center) and the standard deviation (radius). Pooling the data (right) yield a weak but significant support for the phylogenetically least plausible tree (H,G),(L,A). H, G, L, A denote hagfish, lamprey, gnathostomes, and Amphioxus, respectively. (Updated from ref. 82.)

The Four Canonical Gnathostome Clusters

All available evidence supports the hypothesis that all gnathostomes share a system of four *Hox* clusters that arose as a consequence of genome duplications that predate the most recent common ancestor of all recent gnathostomes.[20,90] In particular, the the shark *HoxN* cluster is orthologous to the mammalian *HoxD* cluster, while shark *HoxM* is orthologous to the mammalian *HoxA* cluster.[91,92] Additional support comes from the existence of a shark *HoxB* cluster, whose microRNAs are discussed.[57]

Of course these inferences do not imply that all Gnathostomes have to have at least one homolog for each of the four canonical *Hox* clusters. For instance, even thought there is ample evidence that the shark Heterodontus has orthologs of the *HoxA* and *HoxD* clusters, and a *HoxB* cluster has been isolated, there is no trace of a *HoxC* cluster, in spite of intense efforts (Ruddle and Amemiya, personal communication). The final resolution of that matter will only come with a shark genome sequence, but for now there is no positive evidence for the existence of a *HoxC* cluster in chondrichthyan fishes. The possibility that one of the *Hox* clusters was lost in the shark lineage is not too remote if the *Hox* cluster duplication happened shortly before the bony fish-cartilageous fish split. If there was no strong selection for the diversification of the newly arisen *HoxC* cluster in the shark lineage, the *HoxC* cluster could have been lost without negative consequences.

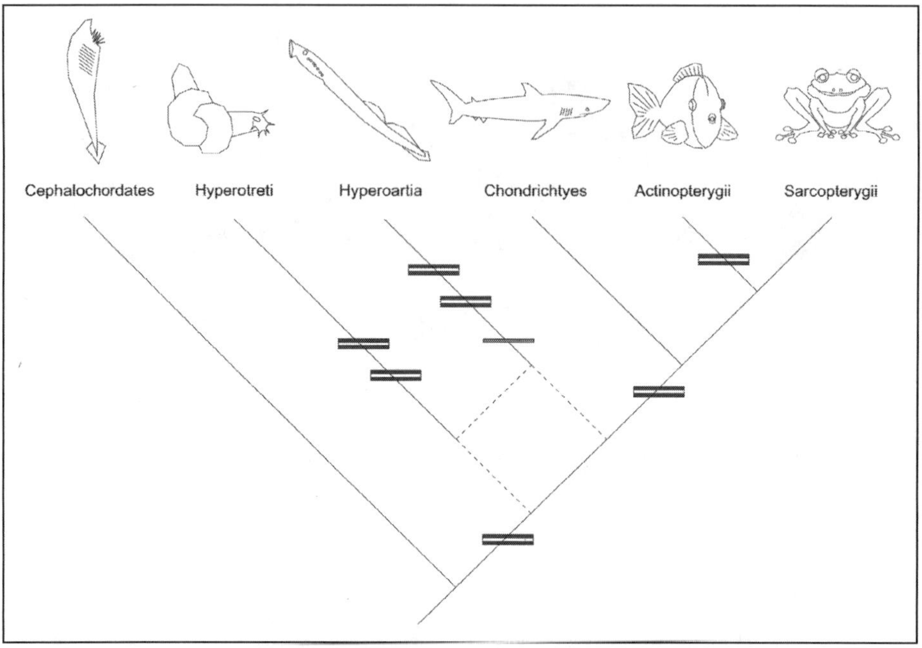

Figure 9. A plausible history of *Hox* cluster duplications that is consistent with the available data. Double bars indicate duplication events, the single bar indicates the loss of one gene in (almost) all paralog groups in the lamprey lineage.

The phylogenetic relationships among the four canonical clusters is hard to determine. Of the three possible un-rooted four taxon trees only the topology can be excluded based on current evidence. Most of the analyses using *Hox* gene sequences alone favor the topology, while the analysis of 5' genes and of collagen genes linked to the *Hox* clusters suggest a topology.[93] The latter analysis clearly showed that the internal branch in either tree is very short but significantly larger than zero, i.e., the likelihood of the star topology is significantly lower than that of topologies with an internal branch. This result suggests that the four clusters arose in two clearly separated duplication events. If duplicated *Hox* genes have a tendency to diverge asymmetrically, as shown for most duplicated genes, long branch attraction would favor the wrong phylogeny, clustering the faster and the slower genes rather than the true first order paralogs.

The Ray-Finned Fish Hox *Cluster Number Expansion*

The first clear evidence that fish (here meaning actinopterygian) *Hox* cluster situation differs from the canonical four clusters came from the discovery of seven *Hox* clusters in the zebrafish.[8] The extent of this phenomenon, however, was not clear for a long time, i.e., whether this was a special feature of or within the cypriniform clade or extended to all ray-finned fishes. Initially the *Hox* cluster complement of Fugu was reported as being a canonical four cluster situation[94] and it was not until 2004 that this was corrected by the work of another lab.[9] It was found that the southern pufferfish *Spheroides nephelus*, as well as the medaka *Oryzias latipes* (see ref. 35 for description of the complete *Hox* cluster), also have seven *Hox* clusters, but the additional clusters are paralogs of the clusters A, B, and D, while in zebrafish they are duplicates of A, B, and C. Gene tree analysis confirmed that the duplicated A and B clusters are ortholog to those of zebrafish and thus that zebrafish and pufferfish are probably derived from a common ancestor which had a duplicated set of eight *Hox* clusters. The differences between zebrafish and pufferfish are thus likely to be due to differential loss of one *HoxD* paralog cluster in the zebrafish lineage and one *HoxC* paralog in the pufferfish lineage after lineage splitting. A more extensive analysis of all *Hox* gene sequences of fishes largely confirmed this picture although the question of whether the *HoxC* genes have been duplicated coincidentally with the *HoxA* and *HoxB* genes was not possible to resolve.[95]

Based on an analysis of syntheny data in the zebrafish and fugu genomes Taylor and collaborators[96] have concluded that the "fish specific" *Hox* cluster duplication occurred in the context of a genome duplication which happened in the ray-finned fish lineage. A molecular clock estimate of the timing of this event suggested an age of 320 Mio years for the fish specific paralogs.[97] Since the timing of the major splits in the fish phylogeny are poorly known (Gauthier, personal communication) this data does not allow us to place the event on the fish phylogeny, i.e., it does not allow us to say how inclusive the clade of ray-finned fishes is that shares these duplicated *Hox* clusters.

A recent gene tree analysis of *Hox* genes from a wider variety of ray-finned fishes showed that the *Hox* cluster duplication was practically coincidental with the initial radiation of the Teleost clade.[11] This confirmed earlier reports which suggested that the *HoxA* clusters of bichir, the most basal ray-finned fish lineage and the paddlefish, member of the second most ancient lineage of ray-finned fish, diverged prior to the duplication of the zebrafish and fugu *HoxA* clusters.[98,99] Ancestral sequence reconstruction further provided evidence that the *Hox* cluster duplication was probably only 7 Mio years prior to the most recent common ancestor of all teleosts, or at most 14 Mio years prior to the crown group teleost node. Hence this evidence shows that the basal ray-finned fishes, like bichir, sturgeons, Amia and gars diverged prior to the *Hox* cluster duplications. The relative dating of duplicated non*Hox* genes also suggests that these originated prior to the teleost radiation and after the divergence of gars and the other basal ray-finned fishes.[100] Taken together these two studies further support the hypothesis that there was a genome duplication coincidental with the radiation of the teleosts which produced the "fish specific" *Hox* clusters.

Overall the data about the duplication of *Hox* genes in the ray-finned fish lineage is the best documented association between a genome duplication event and a major adaptive radiation, which led to the largest clade of recent vertebrates (24,000 species). The timing of the other *Hox* cluster duplication events in early vertebrate history are too poorly constrained to allow inferences whether they are associated with a major evolutionary event.

Recent Duplications

A diverse collection of vertebrate species has undergone recent tetraploidizations, which of course also doubled the *Hox* gene inventory.

Species in the family Salmonidae are believed to have evolved from an ancestor in which an autotetraploidization event occurred 25 to 100 million years ago.[15] Salmonid *Hox* cluster complements seem to be more similar to those of zebrafish (*Danio rerio*) than medaka (*Oryzias latipes*) or pufferfish (*Sphoeroides nephelus* and *Takifugu rubripes*), as both Atlantic salmon and rainbow trout have retained a *HoxCb* ortholog, which has been lost in medaka and pufferfish but not in zebrafish.[16,17]

The amphibian Xenopus is comprised of a series of polyploid species that arose by genome-wide duplication.[101] Most genes, with the exception of certain immune genes, are found in two sets.[102] Differential positive selection in two *HoxB7* paralogs in this lineage is discussed in reference 103.

Several cyprinid lineages are also polypoloid,[104] including goldfish, whose *Hox* genes are discussed in reference 105.

The Effects of Duplication

Gene Loss

Duplicated genes are subsequently lost at a rate between 50 to 70% depending on the study organism (for further references see ref. 109).[106-108] For *Hox* genes, the retention rate

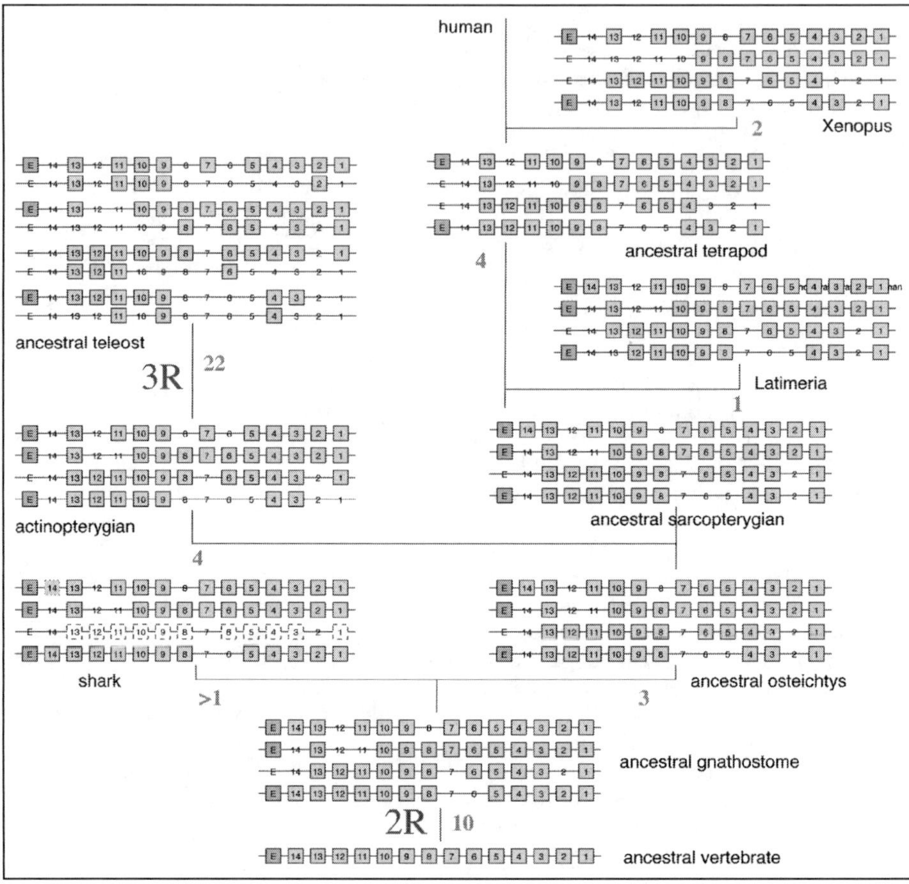

Figure 10. Gene loss in the vertebrate lineage. Data combined from references 17, 112, 113 and references therein. Red number give the number of *Hox* genes lost along each edge. The non*Hox even-skipped* gene is indicated in a different color. Pseudogenes have broken outline. A *HoxC* cluster has not been reported in shark so far. A color version of this figure is available online at www.Eurekah.com.

of duplicated genes highly variable. In between the first and the second duplication 36% of the genes duplicated in the first *Hox* cluster duplication where lost. After the second duplication about 30% of the genes where lost in the mammalian lineage. These numbers are lower than those from other genes cited above. In contrast the "fish specific" *Hox* cluster duplication is associated with a loss of 88% in the zebrafish lineage and 93% in the Fugu lineage.[18] The higher fraction of gene loss in the teleosts can not be explained by different amounts of time that have passed since the duplication, because the "fish specific" duplication is much more recent than the gnathostome, i.e., second duplication. It thus seems that the likelihood of retention of duplicated genes has declined in vertebrate phylogeny from levels above the average of other genes to rates much lower.

The process of reducing the redundancy that arose through the duplication has been very slow and presumably is not yet completed.[110] This view is supported by the existence of a number of easily identifiable *Hox* pseudo-genes. In Figures 10 and 11 we summarize the history of gene losses in the wake of the 2R and 3R genome duplications, respectively. The loss of

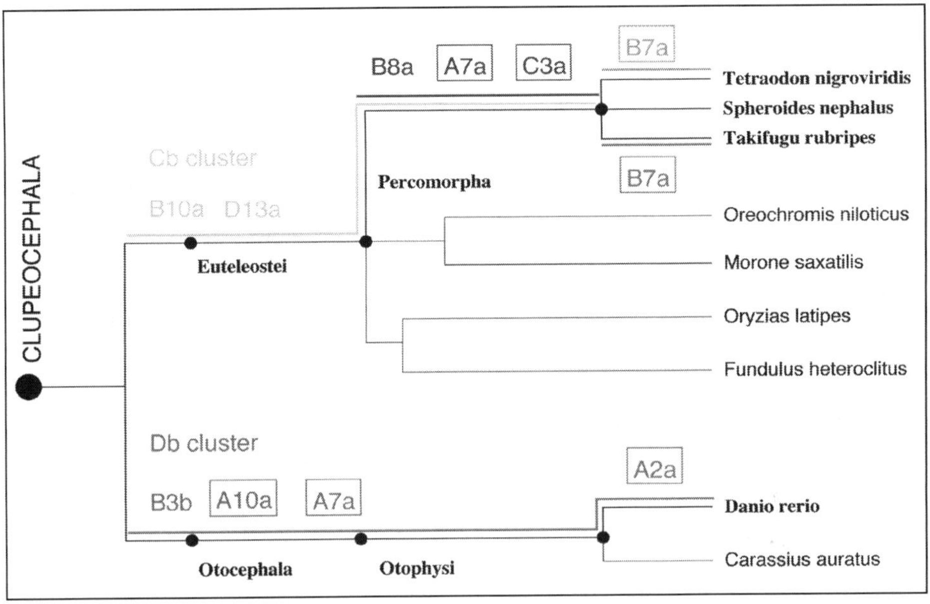

Figure 11. Gene loss in teleost fishes. Teleost fishes for which at least a complete *Hox* gene inventory is known from genomic sequencing are shown in bold font. For the other species given only partial information is available. The loss of *HoxA2a* in zebrafish is very recent, after the divergence of zebrafish and goldfish lineages.[114] Colored lines indicate the maximal evolutionary time period for the loss of genes denoted by the same color. Boxed gene names represent the conversion into pseudo-genes, i.e., a loss event which is still in progress. It seems plausible to assume that the conversion of genes to pseudo-genes occurred close to the leaves of the tree, while gene loss that did not leave detectable traces occurred much earlier. (Updated from ref. 95.) A color version of this figure is available online at www.Eurekah.com.

hoxB7a in some but not all pufferfishes appears be very recent,[110] the conversion of *hoxA7a* into a pseudo-gene independently occurred in zebrafish and the pufferfish lineages (and again in the bichir),[111] while the loss of *hoxB8a* occurred early in part of the percomorpha lineages. Additional recent gene loss has been reported in Lissamphibia.[112]

As discussed above, *Hox* gene loss is in general accompanied by cluster shrinkage and the appearance of repetitive elements at sites of lost genes. This is clearly visible both for the gnathostome clusters following the 2R duplication (Fig. 3) and for the teleost-specific duplication (Fig. 12). For instance, the *HoxAb* and *HoxBb* clusters of the zebrafish show this effect quite dramatically.

Adaptive Evolution in Coding Sequences

The rate of coding sequence evolution in duplicated *Hox* genes of teleosts has been shown to be increased compared to the unduplicated orthologs[115] and there is some evidence that duplicated *Hox* genes experienced directional selection.[116] These findings are consistent with the idea that the duplicated *Hox* genes became involved in adaptive evolutionary changes and played an active role in the evolution of the teleost disparity and diversity.

The ancestral (2R) duplication that gave rise to the paralogous *hoxA7* and *hoxB7* genes followed by a period of time in which *hoxa7* genes diversified under positive selection.[103] The same work shows that one of the *hoxB7* genes that were produced in the Xenopus ancestor by the recent tetraploidization in the amphibian lineage also changed under positive selection.

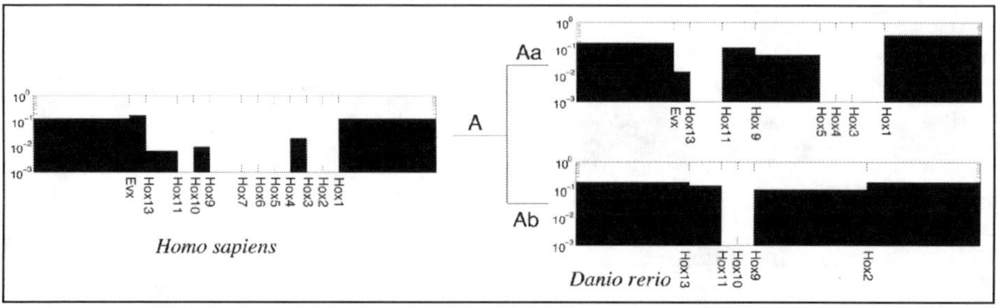

Figure 12. Gene loss after the 3R genome duplication allows repetitive elements to invade the *Hox* clusters, shown here for the unduplicated human *HoxA* cluster and the two corresponding zebrafish paralogs.

The divergence among paralog genes tends to be asymmetric with one paralog diverging faster than the other. In the fugu all the b-paralogs diverge faster than the a-paralogs, while in zebrafish *hoxa13a* diverges faster.[117] It has been reported in reference 118 that such species-specific asymmetric rates of molecular evolution are a relatively common phenomenon in teleosts, which is particularly prevalent in transcription factors.

Conserved Noncoding Sequences

The duplication of the *Hox* cluster is followed not only by gene-loss but also by extensive loss of noncoding sequence conservation.[33] There are three biologically distinct processes that can account for this phenomenon:

1. Structural loss is the loss of putative cis-regulatory elements due to gene loss and/or stochastic resolution of genetic redundancy in the aftermath of the duplication event.
2. Binding site turnover is loss of noncoding sequence conservation due to the replacement of binding sites even though the function of the enhancer remains conserved. This was first documented in the Drosophila *even skipped stripe 2 enhancer*[119] and has since been documented for many other invertebrate taxa. In vertebrates, however, no widespread binding site turnover has been documented.[120]
3. Adaptive modification subsumes changes in the sequence of cis-regulatory sites due to directional natural selection. This effect is thus be associated with functional differences. Drift in subsets of the elements in the regulatory modules is, for example, responsible for the differential expression of *hoxA2a* and *hoxA2b* in *Takifugu rubripes*.[121]

Loss of noncoding sequence conservation is associated with other structural changes, most notably gene loss. Hence the question arises whether the amount of loss observed is more than what should be expected from the changes in the gene-content. This can be addressed by means of a quantitative model[92] that addresses the three main structural causes of footprint loss: (1) Clearly, if a gene is lost, also the associated cis-regulatory elements will be lost, disregarding enhancer sharing. (2) cross-regulatory interactions within the gene cluster may be lost, and (3) there is a loss of enhancers due to stochastic resolution of genetic redundancy. Based on these duplication is a combination of four terms: (I) loss of conserved DNA associated with gene loss, (II) stochastic resolution of redundancy for retained duplicates, (III) loss of conserved DNA associated with the loss of a cross-regulating gene, (IV) nonstructural contributions:

$$r = \underbrace{r(G)}_{(I)} \times \underbrace{\left[\tfrac{1}{2}P_1 + (1 - P_1)\right]}_{(II)} \times \underbrace{(1 - dQ)}_{(III)} \times \underbrace{(1 - \alpha)}_{(IV)}$$

Here, $r(G)$ is the retention rate of genes in focus cluster, P_1 is the fraction of genes with 1st order paralogs, Q is the fraction of extinct genes in entire network, and d is fraction of genes in the cluster from which a gene is cross regulated; for the case of *Hox* clusters we conservatively

Table 1. Loss of conserved noncoding DNA following the teleost-specific duplication of the HoxA cluster

	# Genes	f(G)	P1	Q	r_{obs}	α
DrAa	7	0.63	0.43	5/42	0.31	0.29
DrAb	5	0.45	0.60		0.23	0.18
TrAa	9	0.82	0.56	8/42	0.37	0.22
TrAb	5	0.45	1.00		0.10	**0.48**

Adapted from reference 92.

use the approximation $d \approx 1$ since it is well known that *Hox* genes, for instance, are cross-regulatory, i.e., a *Hox* gene can be the regulatory input for other *Hox* genes.

Applying the structural loss model to the footprint loss data of the *HoxA* clusters shows that the observed amount of retention is in all cases less than predicted as the minimal amount of retention, i.e., the estimated values of α are significantly larger than 0 (Table 1).

Differential Evolution Rates of Noncoding Sequences

The nonstructural modification rate in the fugu *HoxAa* cluster about the same as in zebrafish, while the nonstructural modification rate in the Takifugu *HoxAb* cluster, drastically enhanced relative to zebrafish. Assuming that the probability of functionally conservative binding site turnover is about the same in the two paralog clusters, this result strongly suggests that the Takifugu *HoxAb* cluster experienced adaptive modification at a higher rate than both the Takifugu *HoxAa* cluster and either of the zebrafish clusters.

A relative rate test was developed To be able to rigorously test for asymmetry in the divergence of duplicated conserved noncoding sequences.[122] This approach uses conserved noncoding sequences detected in two outgroup species prior to the duplication and compares the homologous sequences in two paralog clusters and performs a Relative Rate Test on the degree of divergence. This method was applied to the conserved noncoding sequences in the zebrafish and fugu *HoxA*

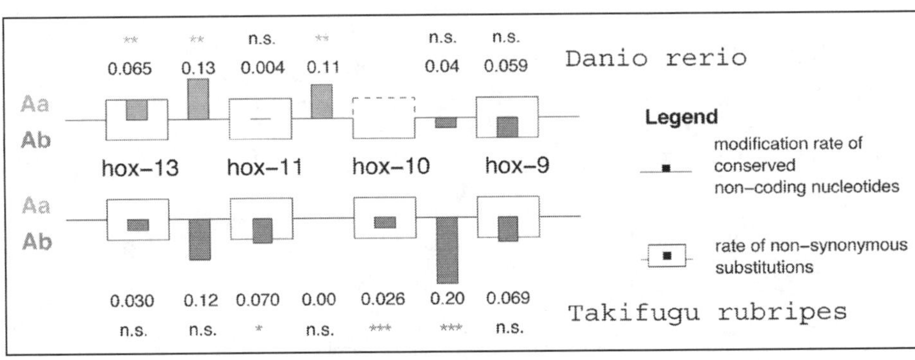

Figure 13. Spatial distribution of sequence divergence along the 5' segment of the duplicated teleost *HoxA* clusters. The open boxes symbolize the coding regions (ignoring introns) and the lines the intergenic sequences (not to scale). The boxes in the intergenic regions give the difference in the fraction of conserved noncoding sequence positions modified in one of the two paralogs:. If the difference is positive, i.e., if the *HoxAa* cluster is more modified than the *HoxAb* cluster the bar is on top of the line, if the *HoxAb* cluster is more modified, then the bar is drawn below the line. Bars in the coding regions symbolize the difference in the rate of nonsynonymous substitutions. Stars indicate statistical significance * ***; n.s.: adapted from reference 117.

clusters.[117] The results showed significant differences in the rate of divergence of conserved noncoding sequences in the zebrafish *hoxA13-hoxA11* and the *hoxA11-hoxA10* intergenic regions (Fig. 13). Interestingly, the direction of the asymmetry is the same as that in the*hoxA13a/b* paralogs, favoring the Aa paralog. In contrast the conserved noncoding sequences of the fugu *HoxAb* cluster diverge more rapidly than the *HoxAa* cluster, also consistent with the asymmetry in the coding regions on the same cluster. Hence there seems to be an association between noncoding and coding region divergence rates, such that the sequences on the same paralog cluster are diverging faster than their counterpart.

Hypotheses and Inferences

Vertebrate *Hox* gene clusters have a tendency to accumulate, but the causes and consequences of this tendency are not clear. This tendency stands in sharp contrast to the evolutionary trends among nonvertebrate metazoans at least among the bilaterian animals. In early metazoan evolution homeobox gene clusters also proliferated (Ferrier this volume). In vertebrates *Hox* cluster duplications occur in the context of whole genome duplications rather than by local tandem duplications. Another special feature of vertebrate *Hox* clusters is that they are structurally much more constrained than their invertebrate counterparts. Structural changes to gnathostome *Hox* clusters are limited to periods following cluster duplication. It is tempting to speculate that there might be a connection between the structural constraints and the tendency to accumulate during vertebrate phylogeny.[18]

One possibility for a connection between the accumulation of vertebrate *Hox* clusters and their structural constraints is that a cluster duplication could temporarily lift the constraints due to functional redundancy and thus open a window of evolvability.[18] If this window of evolvability coincides with an adaptive radiation selection on *Hox* genes and the cluster structure variation might lead to the differentiation among duplicated cluster thus to their maintenance. Consistent with this interpretation is the finding that *Hox* gene paralogs tend to experience directional selection immediately following the duplication event.[11,103,116,123]

An alternative view is that duplicated *Hox* genes undergo passive sub-functionalization due to complementary degeneration of modular functional domains.[124] There is some evidence for subfunctionalization between *hoxA1* and *hoxB1* where *hoxA1* retained inducibility by retinoic acid, and *hoxB1* an autoregulatory element.[125] If sub-functionalization is the main reason for the maintenance of duplicated *Hox* genes in vertebrates, the difference between invertebrates and vertebrates is either due to a lower frequency of genome duplications or a smaller number of sub-functions of invertebrate *Hox* genes (i.e., a lower chance of gene retention due to sub-functionalization) or both. It would be interesting to develop methods to compare the rate of genome duplications in different lineages, in particular between insects and vertebrates.

The considerations about *Hox* clusters and their evolutionary tendencies do not address the question why vertebrates also seem to have more copies of other genes than *Hox* genes. For instance in Drosophila there is one copy of a hedgehog gene (*hh*) but it mammals there are three: *Shh*, *Ihh*, and *Dhh*. One possibility is that in vertebrates lineages with genome duplications tend to be more successful than in invertebrate lineages. Rigorous methods to compare rates of diversification between clades are necessary to address these issues.[10]

References

1. Gehring WJ. Master Control genes in development and evolution: The Homeobox story (Terry Lecture Series). New Haven: Yale University Press, 1998.
2. Cuvier G. Le règne animal; distribué d'après son organisation; pour servir de base à l'histoire naturelle des animaux et d'introduction à l'anatomie comparée. Deterville, Paris: 1817:4.
3. Powers TP, Hogan J, Ke Z et al. Characterization of the hox cluster from the mosquito anopheles gambiae (diptera: Culicidae). Evol Dev 2000; 2(6):311-325.
4. Shashikant CS, Utset MF, Violette TL et al. Homeobox genes in mouse development. Crit Rev Eukaryot Gene Expr 1991; 1:207-245.
5. Pendleton J, Nagai BK, Murtha MT et al. Expansion of the Hox gene family and the evolution of chordates. Proc Natl Acad Sci USA 1993; 90:6300-6304.

6. Murtha MT, Leckman JF, Ruddle FH. Detection of homeobox genes in development and evolution. Proc Natl Acad Sci USA 1991; 88(23):10711-10715.
7. Ferrier DEK, Minguillón C, Holland PWH et al. The amphioxus Hox cluster: Deuterostome posterior flexibility and Hox14. Evol Dev 2000; 2:284-293.
8. Amores A, Force A, Yan YL et al. Postlethwait. Zebrafish hox clusters and vertebrate genome evolution. Science 1998; 282:1711-1714.
9. Amores A, Suzuki T, Yan YL et al. Developmental roles of pufferfish hox clusters and genome evolution in ray-fin fish. Genome Res 2004; 14(1):1-10.
10. Donoghue PCJ, Purnell MA. Genome duplication, extinction, and vertebrate evolution. Trends Ecol Evol 2005; 20:312-319.
11. Crow KD, Stadler PF, Lynch VJ et al. The fish-specific hox cluster duplication is coincident with the origin of teleosts. Mol Biol Evol 2006; 23(1):121-136.
12. Carroll SB, Weatherbee SD, Langeland JA. Homeotic genes and the regulation and evolution of insect wing number. Nature 1995; 375:58-61.
13. Lynch VJ, Roth JJ, Takahashi T et al. Adaptive evolution of hoxa-11 and hoxa-13 at the origin of the uterus in mammals. Proc Biol Sci 2004; 271(1554):2201-2207.
14. Cannatella DC, De Sá RO. Xenopus laevis as a model organism. Syst Biol 1993; 42:476-507.
15. Johnson KR, Wright JE, May B. Linkage relationships reflecting ancestral tetraploidy in salmonid fish. Genetics 1987; 116(4):579-591.
16. Moghadam HK, Ferguson MM, Danzmann RG. Evidence for Hox gene duplication in rainbow trout (Oncorhynchus mykiss): A tetraploid model species. J Mol Evol 2005; 61:804-818.
17. Moghadam HK, Ferguson MM, Danzmann RG. Evolution of Hox clusters in salmonidae: A comparative analysis between atlantic salmon (Salmo salar) and rainbow trout (Oncorhynchus mykiss). J Mol Evol 2005; 61:636-649.
18. Wagner GP, Amemiya C, Ruddle F. Hox cluster duplications and the opportunity for evolutionary novelties. Proc Natl Acad Sci USA 2003; 100:14603-14606.
19. Crow KD, Wagner GP. What is the role of genome duplication in the evolution of complexity and diversity? Mol Biol Evol 2006; 23:887-892.
20. Garcia-Fernández J, Holland PW. Archetypal organization of the amphioxus hox gene cluster. Nature 1994; 370:563-566.
21. Aboobaker AA. Hox gene loss during dynamic evolution of the nematode cluster. Curr Biol 2003; 13:37-40.
22. Aboobaker A, Blaxter M. Hox gene evolution in nematodes: Novelty conserved. Curr Op Genet Devel 2003; 13:593-598.
23. Negre B, Ranz JM, Casals F et al. A new split of the hox gene complex in drosophila: Relocation and evolution of the gene labial. Mol Biol Evol 2003; 20:2042-2054.
24. Negre B, Casillas S, Suzanne M et al. Conservation of regulatory sequences and gene expression patterns in the disintegrating Drosophila Hox gene complex. Genome Res 2005; 15:692-700.
25. Yasukochi Y, Ashakumary LA, Wu C et al. Organization of the Hox gene cluster of the silkworm, Bombyx mori: A split of the Hox cluster in a nonDrosophila insect. Dev Genes Evol 2004; 214:606-614.
26. Pierce RJ, Wu W, Hirai H et al. Evidence for a dispersed Hox gene cluster in the platyhelminth parasite Schistosoma mansoni. Mol Biol Evol 2005; 22:2491-2503.
27. Cameron RA, Rowen L, Nesbitt R et al. Unusual gene order and organization of the sea urchin hox cluster. J Exp Zoolog B (Mol Dev Evol) 2006; 306:45-58.
28. Spagnuolo A, Ristoratore F, Di Gregorio A et al. Unusual number and genomic organization of Hox genes in the tunicate Ciona intestinalis. Gene 2003; 309:71-79.
29. Ikuta T, Yoshida N, Satoh N et al. Ciona intestinalis hox gene cluster: Its dispersed structure and residual colinear expression in development. Proc Natl Acad Sci USA 2004; 101:15118-15123.
30. Ikuta T, Saiga H. Organization of hox genes in ascidians: Present, past, and future. Dev Dyn 2005; 233:382-389.
31. Seo HC, Edvardsen RB, Maeland AD et al. Hox cluster disintegration with persistent anteroposterior order of expression in Oikopleura dioica. Nature 2004; 431:67-71.
32. Edvardsen RB, Seo HC, Jensen MFJ et al. Remodelling of the homeobox gene complement in the tunicate Oikopleura dioica. Curr Biol 2005; 15:R12-R13.
33. Chiu CH, Amemiya C, Dewar K et al. Molecular evolution of the HoxA cluster in the three major gnathostome lineages. Proc Natl Acad Sci USA 2002; 99:5492-5497.
34. Santini S, Boore JL, Meyer A. Evolutionary conservation of regulatory elements in vertebrate Hox gene clusters. Genome Res 2003; 13:1111-1122.
35. Kurosawa G, Takamatsu N, Takahashi M et al. Organization and structure of hox gene loci in medaka genome and comparison with those of pufferfish and zebrafish genomes. Gene 2006; 370:75-82.

36. Lee AP, Koh EG, Tay A et al. Highly conserved syntenic blocks at the vertebrate Hox loci and conserved regulatory elements within and outside hox gene clusters. Proc Natl Acad Sci USA 2006; 103:6994-6999.
37. Fried C, Prohaska SJ, Stadler PF. Exclusion of repetitive DNA elements from gnathostome Hox clusters. J Exp Zool Mol Dev Evol 2004; 302B:165-173.
38. Casals F, Caceres M, Ruiz A. The foldback-like transposon Galileo is involved in the generation of two different natural chromosomal inversions of Drosophila buzzatii. Mol Biol Evol 2003; 20:674-685.
39. Lewis EB, Pfeiffer BD, Mathog DR et al. Evolution of the homeobox complex in Diptera. Curr Biol 2003; R587-R588.
40. Britten RJ. DNA sequence insertion and evolutionary variation in gene regulation. Proc Natl Acad Sci USA 1996; 93:9374-9377.
41. Stenger JE, Lobachev KS, Gordenin D et al. Biased distribution of inverted and direct Alus in the human genome: Implications for insertion, exclusion, and genome stability. Genome Res 2001; 11:12-27.
42. Okazaki Y, Furuno M, Kasukawa T et al. Analysis of the mouse transcriptome based on functional annotation of 60,770 full-length cDNAs. Nature 2002; 420:563-573.
43. Imanishi T, Itoh T, Suzuki Y et al. Integrative annotation of 21,037 human genes validated by full-length cDNA clones. PLoS Biology 2004; 2:0856-0875.
44. Bertone P, Stoc V, Royce TE et al. Snyder. Global identification of human transcribed sequences with genome tiling arrays. Science 2004; 306:2242-2246.
45. Kampa D, Cheng J, Kapranov P et al. Novel RNAs identified from an in-depth analysis of the transcriptome of human chromosomes 21 and 22. Genome Res 2004; 14:331-342.
46. Johnson JM, Edwards S, Shoemaker D et al. Dark matter in the genome: Evidence of widespread transcription detected by microarray tiling experiments. Trends Genet 2005; 21:93-102.
47. Cheng J, Kapranov P, Drenkow J et al. Transcriptional maps of 10 human chromosomes at 5-nucleotide resolution. Science 2005; 308:1149-1154.
48. Washietl S, Hofacker IL, Lukasser M et al. Mapping of conserved RNA secondary structures predicts thousands of functional noncoding RNAs in the human genome. Nature Biotech 2005; 23:1383-1390.
49. Pedersen JS, Bejerano G, Siepel A et al. Identification and classification of conserved RNA secondary structures in the human genome. PLoS Comput Biol 2006; 2:e33.
50. Benson DA, Karsch-Mizrachi I, Lipman D et al. GenBank: Update. Nucl Acids Res 2004; 32:D23-26.
51. Sugnet CW, Kent WJ, MA et al. Transcriptome and genome conservation of alternative splicing events in humans and mice. Pacific Symp Biocomp 2004; 9:66-77.
52. RIKEN Genome Exploration Research Group. Antisense transcription in the mammalian transcriptome. Science 2005; 309:1564-1566.
53. Potter SS, Brandford WW. Evolutionary conservation and tissue-specific processing of Hoxa 11 antisense transcripts. Mamm Genome 1998; 9:799-806.
54. Michael Chau SPY, Taylor HS. HOXA11 silencing and endogenous HOXA11 antisense ribonucleic acid in the uterine endometrium. J Clin Endocrin Metabolism 2002; 87:2674-2680.
55. Liu XF, Olsson P, Wolfgang CD et al. PRAC: A novel small nuclear protein that is specifically expressed in human prostate and colon. Prostate 2001; 47:125-131.
56. OlssonP, Motegi A, Bera TK et al. PRAC2: A new gene expressed in human prostate and prostate cancer. Prostate 2003; 56:123-130.
57. Tanzer A, Amemiya CT, Kim CB et al. Evolution of microRNAs located within Hox gene clusters. J Exp Zool Mol Dev Evol 2005; 304B:75-85.
58. Chopra VS, Mishra RK. "mir"acles in Hox gene regulation. Bioessays 2006; 28:445-448.
59. Hadrys T, Prince V, Hunter M et al. Comparative genomic analysis of vertebrate Hox3 and Hox4 genes. J Exp Zool B Mol Dev Evol 2004; 302:147-164.
60. Yekta S, Shih IH, Bartel DP. MircoRNA-directed cleavage of HoxB8 mRNA. Science 2004; 304:594-596.
61. Irvine SQ, Carr JL, Bailey WJ et al. Genomic analysis of Hox clusters in the sea lamprey, Petromyzon marinus. J Exp Zool (Mol Dev Evol) 2002; 294:47-62.
62. Ronshaugen M, Biemar F, Piel J et al. The drosophila microRNA iab-4 causes a dominant homeotic transformation of halteres to wings. Genes Dev 2005; 19:2947-2952.
63. Delsuc F, Brinkmann H, Chourrout D et al. Tunicates and not cephalochordates are the closest living relatives of vertebrates. Nature 2006; 439(7079):923-924.
64. Campos PRA, de Olivera VM, Wagner GP et al. Gene phylogenies and protein-protein interactions: Possible artifacts resulting from shared protein interaction partners. J Theor Biol 2004; 231:197-202.

65. Misof BY, Wagner GP. Evidence for four hox clusters in the killifish fundulus heteroclitus (teleostei). Mol Phylogenet Evol 1996; 5(2):309-322.
66. Misof BY, Blanco MJ, Wagner GP. PCR-survey of hox-genes of the zebrafish: New sequence information and evolutionary implications. J Exp Zool 1996; 274(3):193-206.
67. Powers TP, Amemiya CT. Evidence for a hox14 paralog group in vertebrates. Current Biol 2004; 14:R183-R184.
68. Janvier P. Early Vertebrates. Oxford: Clarendon Press, 1996.
69. Nelson JS. Fishes of the World. New York: John Wiley and Sons Inc., 1994.
70. Forey P, Janvier P. Agnathans and the origin of jawed vertebrates. Nature 1993; 361:129-134.
71. Maisey JG. Heads and tails: A chordate phylogeny. Cladistics 1986; 2:201-256.
72. Furlong RF , Holland PW. Bayesian phylogenetic analysis supports monophyly of ambulacria and cyclostomes. Zool Sci 2002; 19:593-599.
73. Kuraku S, Hoshiyama K, Katoh D et al. Monophyly of lampreys and hagfishes supported by nuclear DNA-coded sequences. J Mol Evol 1999; 49:729-735.
74. Mallatt J, Sullivan J. 28S and 18S rDNA sequences support the monophyly of lampreys and hagfishes. Mol Biol Evol 1998; 15:1706-1718.
75. Stock DW, Whitt GS. Evidence from 18S ribosomal RNA sequences that lampreys and hagfish form a natural group. Science 1992; 257:787-789.
76. Delabre C, Gallut C, Barriel V et al. Complete mitochondrialDNA of the hagfish, Eptatretus burgerie: The comparative anaylsis of mitochondrial DNA sequences strongly supports cyclostome monophyly. Mol Phylog Evol 2002; 22:184-192.
77. Takezaki N, Figueroa F, Zelska-Rutcynska Z et al. Molecular phylogeny of early vertebrates: Monophyly of the agnathans as revealed by sequences of 35 genes. Mol Biol Evol 2003; 20:287-292.
78. Gürsoy HC, Koper D, Benecke BJ. The vertebrate 7S K RNA separates hagfish (Myxine glutinosa) and lamprey (Lampetra fluviatilis). J Mol Evol 2000; 50:456-464.
79. Rasmussen AS, Janke A, Arnason U. The mitochondrial DNA molecule of the hagfish Myxine glutinosa and vertebrate phylogeny. J Mol Evol 1998; 46:382-388.
80. Sharman AC, Holland PW. Estimation of Hox gene cluster number in lampreys. Int J Dev Biol 1998; 42:617-620.
81. Force A, Amores A, Postlethwait JH. Hox cluster organization in the jawless vertebrate Petromyzon marinus. J Exp Zool (Mol Dev Evol) 2002; 294:30-46.
82. Fried C, Prohaska SJ, Stadler PF. Independent Hox-cluster duplications in lampreys. J Exp Zool Mol Dev Evol 2003; 299B:18-25.
83. Takio Y, Pasqualetti M, Kuraku S et al. Evolutionary biology: Lamprey hox genes and the evolution of jaws. Nature 2004; 429:262f.
84. Cohn MJ. Evolutionary biology: Lamprey Hox genes and the origin of jaws. Nature 2002; 416:386-387.
85. Escriva H, Manzon L, Youson J et al. Analysis of lamprey and hagfish genes reveals a complex history of gene duplications during early vertebrate evolution. Mol Biol Evol 2002; 19:1440-1450.
86. Neidert AH, VirupannavarV, Hooker GW et al. Lamprey dlx genes and early vertebrate evolution. Proc Natl Acad Sci USA 2001; 98:1665-1670.
87. Germot A, Lecointre G, Plouhinec JL et al. Structural evolution of otx genes in craniates. Mol Biol Evol 2001; 18:1668-1678.
88. Stadler PF, Fried C, Prohaska SJ et al. Evidence for independent Hox gene duplications in the hagfish lineage: A PCR-based gene inventory of Eptatretus stoutii. Mol Phylog Evol 2004; 32:686-692.
89. Nieselt-Struwe K, von Haeseler A. Quartet-mapping, a generalization of the likelihood mapping procedure. Mol Biol Evol 2001; 18:1204-1219.
90. Holland PWH, Garcia-Fernández J, Williams NA et al. Gene duplication and the origins of vertebrate development. Development (Suppl.) 1994:125-133.
91. Kim CB, Amemiya C, Bailey W et al. Hox cluster genomics in the horn shark, Heterodontus francisci. Proc Natl Acad Sci USA 2000; 97:1655-1660.
92. Prohaska SJ, Fried C, Amemiya CT et al. The shark HoxN cluster is homologous to the human HoxD cluster. J Mol Evol 2004; 58:212-217.
93. Bailey W, Kim J, Wagner G et al. Phylogenetic reconstruction of vertebrate hox cluster duplications. Mol Biol Evol 1997; 14(8):843-853.
94. Aparicio S, Hawker K, Cottage A et al. Organization of the Fugu rubripes Hox clusters: Evidence for continuing evolution of vertebrate Hox complexes. Nat Genetics 1997; 16:79-83.
95. Prohaska S, Stadler PF. The duplication of the hox gene clusters in teleost fishes. Th Biosci 2004; 123:33-68.
96. Taylor J, Braasch I, Frickey T et al. Genome duplication, a trait shared by 22,000 species of ray-finned fish. Genome Res 2003; 13:382-390.

97. Vandepoele K, De Vos W, Taylor JS et al. Major events in the genome evolution of vertebrates: Paranome age and size differ considerably between ray-finned fishes and land vertebrates. Proc Natl Acad Sci USA 2004; 101:1638-1643.
98. Chiu CH, Dewar K, Wagner GP et al. Amemiya. Bichir HoxA cluster sequence reveals surprising trends in rayfinned fish genomic evolution. Genome Res 2004; 14:11-17.
99. Metscher BD, Takahashi K, Crow K et al. Expression of hoxa-11 and hoxa-13 in the pectoral fin of a basal ray-finned fish, polyodon spathula: Implications for the origin of tetrapod limbs. Evol Dev 2005; 7:186-195.
100. Hoegg S, Brinkmann H, Taylor J et al. Phylogenetic timing of the fish-specific genome duplication correlates with the diversification of teleost fish. J Mol Evol 2004; 59(2):190-203.
101. Kobel H, Du Pasquier L. Genetics of polyploid xenopus. Trends Genet 1986; 12:310-315.
102. Courtet M, Flajnik M, Du Pasquier L. Major histocompatibility complex and immunoglobulin loci visualized by in situ hybridization on Xenopus chromosomes. Dev Comp Immunol 2001; 25.
103. Fares MA, Bezemer D, Moya A et al. Selection on coding regions determined Hox7 genes evolution. Mol Biol Evol 2003; 20(12):2104-2112.
104. Larhammar D, Risinger C. Molecular genetic aspects of tetraploidy in the common carp Cyprinus carpio. Mol Phylogenet Evol 1994; 3:59-68.
105. Levine EM, Schechter N. Homeobox genes expressed in the retina and brain of adult goldfish. Proc Natl Acad Sci USA 1993; 90:2729-2733.
106. Ferris SD, Whitt GS. Evolution of the differential regulation of duplicate genes after polyploidization. J Mol Evol 1979; 12:267-317.
107. Nadeau JH, Sankoff D. Comparable rates of gene loss and functional divergence after genome duplications early in vertebrate evolution. Genetics 1997; 147:1259-1266.
108. Wendel JF. Genome evolution in polypoids. Plant Mol Biol 2000; 42:225-249.
109. Lynch M, Conery JS. The evolutionary fate and consequences of duplicate genes. Science 2000; 290:1151-1155.
110. Amores A, Suzuki T, Yan YL et al. Developmental roles of pufferfish Hox clusters and genome evolution in ray-fin fish. Genome Res 2004; 14:1-10.
111. Chiu CH, Dewar K, Wagner GP et al. Bichir HoxA cluster sequence reveals surprising trends in ray-finned fish genomic evolution. Genome Res 2004; 14:11-17.
112. Mannaert A, Roelants K, Bossuyt F et al. A PCR survey for posterior hox genes in amphibians. Mol Phylogenet Evol 2006; 38:449-458.
113. Hoegg S, Meyer A. Hox clusters as models for vertebrate genome evolution. Trends Genet 2005; 21:421-424.
114. Luo J, Stadler PF, Meyer A et al. PCR survey of Hox genes in the goldfish. Carassius auratus auratus. 2006, (submitted).
115. Chiu CH, Nonaka D, Xue L et al. Evolution of Hoxa-11 in lineages phylogenetically positioned along the fin-limb transition. Mol Phylogen Evol 2000; 17:305-316.
116. van de Peer Y, Taylor JS, BI et al. The ghost of selection past: Rates of evolution and functinal divergence of anciently duplicated genes. J Mol Evol 2001; 53:436-446.
117. Wagner GP, Takahashi K, Lynch V et al. Molecular evolution of duplicated ray finned fisch hoxa clusters: Increased synonymous substitution rate and asymmetrical codivergence of coding and noncoding sequences. J Mol Evol 2005; 60:665-676.
118. Steinke D, Salzburger W, Braasch I et al. Many genes in fish have species-specific asymmetric rates of molecular evolution. BMC Genomics 2006; 7:20, [epub].
119. Ludwig MZ, Bergman C, Patel NH et al. Evidence for stabilizing selection in a eukaryotic enhancer element. Nature 2000; 403:564-567.
120. Carter AJ, Wagner GP. Evolution of functionally conserved enhancers can be accelerated in large populations: A population-genetic model. Proc R Soc Lond B Biol Sci 2002; 269:953-960.
121. Tumpel S, Cambronero F, Wiedemann LM et al. Evolution of cis elements in the differential expression of two Hoxa2 coparalogous genes in pufferfish (Takifugu rubripes). Proc Natl Acad Sci USA 2006; 103:5419-5424.
122. Wagner G, Fried C, Prohaska S et al. Divergence of conserved noncoding sequences: Rate estimates and relative rate tests. Mol Biol Evol 2004; 21(11):2116-2121.
123. Lynch VJ, Roth JJ, Wagner GP. Adaptive evolution of Hox-gene homeodomains after cluster duplication. 2006, (submitted).
124. Force A, Lynch M, Pickett F et al. Preservation of duplicate genes by complementary, degenerative mutations. Genetics 1999; 151(4):1531-1545.
125. Prince V, Pickett FB. Splitting pairs: The diverging fates of duplicated genes. Nat Rev Genet 2002; 3.

CHAPTER 6

Chromatin and the Control of *Hox* Gene Expression

Laila Kobrossy and Mark Featherstone*

Introduction

Antero-posterior patterning of the animal embryo is governed in part by the highly conserved *Hox* genes. In most animals studied to date, *Hox* genes are assembled within one or more clusters.[1-5] The thirty-nine *Hox* genes of mice and humans are organized into four clusters, each located on a different chromosome.[4] There are thirteen possible gene positions in each cluster, although none of the clusters retains all thirteen members (Fig. 1). *Hox* genes occupying the same relative position between clusters are termed paralogs, sharing high sequence identity and functional redundancy. Because all genes are transcribed in the same direction, one can assign a 3' and a 5' end to a cluster.

Consistent with their key role in embryonic patterning, *Hox* gene expression is tightly regulated at the transcriptional level,[6] and the desire to define the mechanisms controlling this expression has driven many research programs. Consequent to shifts in our understanding of transcriptional regulation generally, these programs have broadened to include not only the identification of sequence-specific transcription factors and the cis-elements to which they bind, but also the role of chromatin modification and remodeling. A detailed introduction to the mechanisms modulating chromatin structure and function is beyond the scope of this discussion, and readers are directed to numerous reviews on the subject.[7-9]

Briefly, the nucleosomal organization of DNA is used by the cell to control transcription at all levels. The chromatin of transcriptionally active loci is described as "open" as a result of its relative accessibility to chemical and enzymatic probes, the relative paucity of nucleosomes, and by its extended appearance. By contrast, the "closed" chromatin of inactive genes is compacted and inaccessible by the same assays. The conversion of chromatin from one state to the other involves two broad classes of enzymes: histone modifying enzymes and chromatin remodelers. The list of histone modifiers and modifications is growing, and includes enzymes with the capacity to add phosphate, methyl and acetyl groups to residues largely in the histone N-terminal tails. Whereas a number of transcriptional coactivators have histone acetyltransferase (HAT) activity, corepressors often harbour histone deacetylase (HDAC) function.[10,11] The precise modifications and sites of modification provide a "histone code"[12] which serves as a beacon for the differential recruitment of specific chromatin remodeling enzymes which then act to open or close chromatin, thus conferring greater or lesser accessibility to components of the transcriptional machinery.

Numerous studies have established that it is the strictly defined anterior expression border that most determines HOX activity, and shifting this border either anteriorly or posteriorly

*Corresponding Author: M. Featherstone—School of Biological Sciences, Nanyang Technological University, Singapore. Email: msfeatherstone@ntu.edu.sg

HOX Gene Expression, edited by Spyros Papageorgiou. ©2007 Landes Bioscience and Springer Science+Business Media.

leads to embryonic malformations and homeotic transformations.[13,14] A striking feature of transcriptional activation within *Hox* clusters is a process termed "collinearity," in which both the temporal and spatial order of activation is correlated with the relative position of a particular *Hox* gene within a cluster.[15] Thus, in general, genes located more 3' are expressed earlier and have a more anterior expression border than genes located more 5' along the cluster (Fig. 1). This and other observations have implied a sequential opening of chromatin in a given Hox complex, starting at the 3' end of a cluster and moving successively 5'.[13,16-20]

In this review, we discuss recent insights into the role of chromatin in vertebrate *Hox* gene regulation, focusing on transcriptional initiation and maintenance.

Anticipation

Accumulating evidence suggests that the modulation of chromatin structure at Hox loci begins prior to transcriptional initiation. Histone modifications indicative of open chromatin at the murine *Hoxd4* locus are observed early on in posterior embryonic compartments that will later express *Hoxd4*.[19] A comparable observation has been made for genes of the HoxB cluster in retinoic-acid-treated embryonic stem cells which activate Hox gene expression in a temporally collinear fashion.[21] Thus, chromatin modification anticipates Hox gene activation.

A different approach suggests that 3' *Hox* genes, at which collinear activation begins, are transcriptionally available prior to the actual initiation of transcription. When early embryos (E6.0-6.2) not yet expressing HoxB genes are cultured in the presence of RA, the expression of 3' genes (*Hoxb1* and *Hoxb2*) is precociously activated, whereas that of more 5' genes (*Hoxb3*, *Hoxb4*, *Hoxb6*, *Hoxb8*) is not. Moreover, a *Hoxb1-lacZ* transgene inserted randomly within the genome does not respond to early RA treatment, unlike its endogenous, cluster-embedded counterpart. These results suggest that the early transcriptional availability of *Hoxb1* is a cluster-specific property which could be accounted for by chromatin modification or remodeling events that anticipate and facilitate subsequent initiation events.[17]

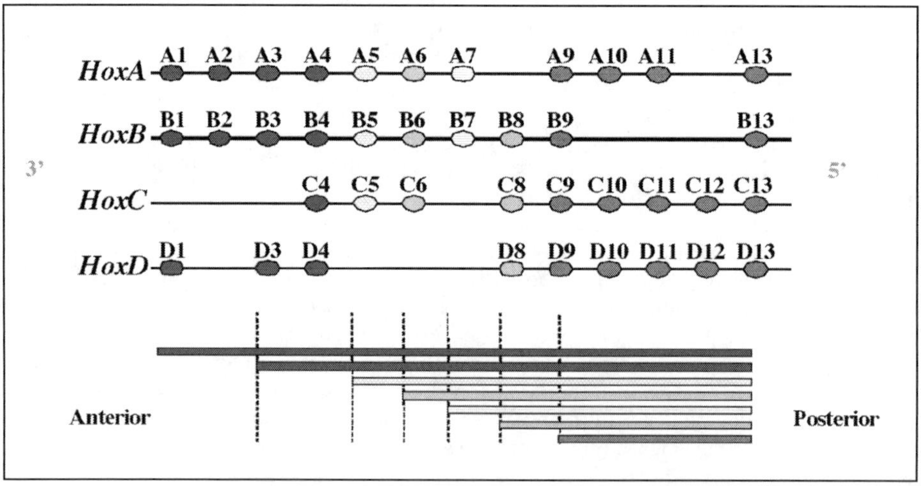

Figure 1. Genomic organization of the mammalian *Hox* complex. Taken together, the four Hox clusters reveal 13 potential positions for Hox genes, though no single cluster has genes at all 13 positions. There are therefore 13 paralog groups, each comprising genes at the same relative position between clusters. The 3' and 5' ends represent the direction of transcription. Coloured lines below the clusters represent the relative expression domain for successive groups of Hox genes along the AP axis of the trunk. Note that in reality each paralog group (eg. group1, comprising *Hoxa1*, *Hoxb1* and *Hoxd1*) has a distinct anterior border. Within paralog groups anterior borders can also differ.

The mechanism by which anticipation is accomplished is unknown, but could involve the prior binding of so-called "pioneer" proteins to closed chromatin.[22] The pioneer then marks chromatin for subsequent opening by recruitment of later-appearing transcription factors. Such a role has been described for PBX1 which recruits MYOD to the silent myogenin gene.[23] We can speculate that the retinoid receptors, already so implicated in Hox gene activation (see below), could act as pioneers given their known ability to bind active and inactive loci.[24] Nonexclusively, anticipation may invoke global enhancers positioned to either side of the Hox cluster such as has been described for regulatory regions directing collinear HoxD expression in the forearm and digits of the developing limbs.[13,25]

Initiation

Retinoic Acid

Retinoic acid (RA) regulates gene expression by binding to the ligand-dependent RA receptors, sequence-specific DNA-binding transcription factors which orchestrate the modification of chromatin structure via coactivator and corepressor recruitment. There are two classes of RA receptors: RARs, which bind both all-trans and 9-cis RA, and RXRs which only bind 9-cis RA. RARs and RXRs bind as heterodimers to a site known as a retinoic acid response element (RARE). In the absence of RA, RAR/RXR heterodimers recruit corepressors such as NCoRs, and HDACs resulting in a transcriptionally inaccessible chromatin structure.[24] However, binding to RA results in an allosteric change in RAR/RXR ligand binding domains, and recruitment of coactivators bearing HAT activity such as CBP/p300.[26] Functional retinoic acid response elements (RAREs), the binding sites for retinoid receptors, have been identified for *Hoxa1*, *Hoxb1*, *Hoxa4*, *Hoxb4* and *Hoxd4*.[27-34] In addition, an RARE discovered between *Hoxb4* and *Hoxb5* directs late hindbrain expression of *Hoxb5*, *Hoxb6* and *Hoxb8*.[35] A negative role for unliganded RAR-RXR in Hox expression is less clear, but at least one repressive RARE has been described in a Hox regulatory region.[33]

When a *Hoxd11-lacZ* transgene is inserted upstream of *Hoxd13*, the most 5' gene of the HoxD cluster (Fig. 1), transcription of the transgene in the limb bud is delayed relative to endogenous *Hoxd11*. Expression first resembles that of *Hoxd13*, but later regains characteristics of endogenous *Hoxd11*, attributing the delay to a misregulation of transcriptional initiation.[36] Similar results were obtained for a *Hoxd9* transgene inserted 5' to its endogenous position, suggesting the existence of a higher order regulatory mechanism which controls and inhibits premature transcriptional initiation of 5' *Hox* genes. This regulatory pathway would ensure that genes located more 3' along the cluster are available for transcriptional initiation earlier than more 5' genes, probably due to chromatin opening or modification at the 3' end, and/or a silencing mechanism localized at the 5' end of the cluster.[13,25,37]

A role for chromatin remodeling in mediating the transcriptional repression of 5' *HoxD* genes was described following targeted deletion of the gene encoding murine zinc-finger protein PLZF (promyelocytic leukemia zinc finger).[38] *Plzf*$^{-/-}$ mice exhibit skeletal and limb patterning defects as well as anteriorization of several 5' *HoxD* genes. This study demonstrated that PLZF homodimers mediate transcriptional repression by facilitating the formation of DNA loops by bridging distant PLZF binding sites. PLZF then recruites HDACs and the PcG protein BMI1 which results in histone deacetylation and transcriptional silencing (see below), as demonstrated by decreased activity of a *Hoxd11* reporter construct.

Recent studies have tracked the physical changes in chromatin structure and nuclear organization at the *HoxB* cluster following gene activation.[21,39] Treating embryonic stem (ES) cells with RA results in a sequential activation of *HoxB* genes starting at the 3' end.[21] This is accompanied by a visible decondensation of chromatin at the *HoxB* locus and a progressive extrusion of genes out of their chromosomal territories (CT) which starts at the 3' end and corresponds to the *Hox* gene being expressed during that timeframe. The same group later showed that similar chromatin changes take place at the *HoxB* cluster during different stages of embryogenesis.[39]

During gastrulation, the expression of *Hoxb1* in the primitive streak is accompanied by chromatin opening and the extrusion of the *Hoxb1* locus from its CT as shown by fluorescent in situ hybridization (FISH). These changes are also true in the hindbrain at later stages for cells expressing *Hoxb1* in rhombomere (r) 4, but are absent from cells in r1, r2 and r5 where *Hoxb1* is not expressed. *Hoxb9* moves out of its CT only at later stages and only in the spinal cord where it is being expressed. Chromatin decondensation and changes in CT locations of both *Hoxb1* and *Hoxb9* occurs simultaneously with transcriptional activation, suggesting a developmental mechanism regulating the collinear expression pattern of *Hox* gene expression in the CNS.

Hoxd4

Chromatin modification and remodeling have been correlated with transcriptional initiation at specific *Hox* loci. Sequential 3' to 5' chromatin opening accompanies transcriptional activation of murine *Hoxd4* in the developing embryo and RA-treated P19 embryonal carcinoma cells.[19] Chromatin immunoprecipitation (ChIP) assays using antibodies against RNA polymerase II (RNA Pol II) and markers of open chromatin (acetylation of histone H3 on lysine (K) 9 and K14, and on H4 on K4/7/11/15, and methylation of H3 on K4) reveal that chromatin modification initiates at the 3' neural enhancer before becoming detectable at the promoter upstream. During *Hoxd4* activation in neurally differentiating P19 cells, these histone modifications spread 5' from the enhancer to encompass intermediate sequences before reaching the promoter. This directional chromatin opening reflects the normal 3' to 5' collinear Hox activation across the cluster, however it is not yet known whether it is a part of a cluster-wide mechanism, or simply a local event necessitated by the 3' placement of the *Hoxd4* neural enhancer.

At least three positioned nucleosomes span the *Hoxd4* transcriptional start site, and are subsequently remodeled into a more relaxed conformation during transcriptional activation.[40] This remodeling, however, is not accompanied by nucleosome sliding. SWI/SNF and ISWI are the two major ATP-dependent chromatin remodeling complexes, but only the former can reconfigure nucleosomal DNA in the absence of sliding, implicating SWI/SNF specifically in *Hoxd4* promoter activation.[41]

Interestingly, a functional binding site for the Polycomb group (PcG) protein YY1 maps to an internucleosomal region at the *Hoxd4* promoter.[40] As confirmed by ChIP, YY1 is bound in vivo both before and after *Hoxd4* activation. Reducing the levels of YY1 by siRNA-mediated knockdown activates *Hoxd4* transcription in undifferentiated P19 embryonal carcinoma cells, in line with the known role of PcG proteins in maintaining Hox genes in a silent state. However, in RA-treated P19 cells which now express *Hoxd4*, YY1 results in a further increase in the levels of *Hoxd4* mRNA. These results strongly suggest that YY1 negatively regulates *Hoxd4* both before and after transcriptional initiation, ablating expression in the former, while fine tuning in the latter. This quantitative shift in YY1 function may be regulated by association with MEL18 which is present with YY1 at the promoter before, but not after, initiation. It may also be speculated that YY1 embedded in the closed chromatin of the *Hoxd4* promoter could carry out pioneer functions as described above. The findings for *Hoxd4* further raise the possibility that other Hox genes regulated by YY1 may be similarly modulated.[42,43]

ChIP experiments with *Hoxd4* in E8.0 and E10.5 mouse embryos reveal that differential chromatin modifications at this locus distinguish anterior from posterior compartments. At E8.0, a few hours before *Hoxd4* expression is initiated, various histone modifications specific for open chromatin, as well as RNA Pol II, are detected at 3' regulatory regions only in posterior embryonic compartments, where *Hoxd4* is later expressed.[19] These modifications are not detected in tissue anterior to r7 where *Hoxd4* is never expressed. By E10.5, markers of relaxed chromatin have encompassed the entire locus in spinal cord cells which express *Hoxd4*. By contrast, H3/H4 acetylation, K4 methylation (a marker of open chromatin), and RNA Pol II are still not detected at *Hoxd4* in anterior, nonexpressing tissue. Following from these results, it seems likely that the cellular behaviour classically described by the terms commitment and determination have their molecular underpinning in the distinct chromatin states of key loci which are more (committed) or less (determined) susceptible to resculpting.

Maintenance

In addition to its role in regulating *Hox* transcriptional initiation, modification of chromatin at *Hox* loci is implicated in the maintenance of *Hox* transcriptional status. A global analysis of histone modifications at several *Hox* clusters in human and mouse fibroblasts revealed unique histone methylation and acetylation patterns.[44] Acetylation of histone H3 at K9 and K14 is a marker for open transcriptionally active chromatin, and histone H3 K4 methylation is a marker of active transcriptional elongation as well as transcriptional activation.[12] In this study, broad areas encompassing highly conserved regions of mouse and human Hox sequences (including regulatory and intergenic sequences) were shown to be highly methylated and acetylated. Interestingly, these modification patterns are present only in cell lineages where *Hox* genes are known to be active (lung), and are absent in other cell types (gum, foreskin) which do not express these genes. First, the results suggest that chromatin modifications at *Hox* intergenic sequences are as important as those at other regulatory sequences such as enhancers and promoters. Moreover, the increased H3 K4 methylation patterns might reflect regions which are bound by the TrxG protein MLL (mixed lineage leukemia), a histone methyltransferase which is essential for maintenance of Hox gene expression.

The first observations linking *Hox* gene function with the PcG and Trithorax Group (TrxG) proteins were made in *Drosophila*, when homeotic transformations similar to those provoked by *Hox* gene mutation were linked to nonHox loci. Further genetic and molecular studies identified these gene products as members of the PcG and TrxG family which were later found to play antagonistic roles in maintaining the transcriptional status of *Hox* and other important developmental regulators through the preservation of chromatin states.[45] PcG and TrxG homologs have since been characterized in mammals where accumulating evidence suggests a highly conserved role in Hox regulation here as well.

PcG Proteins

In general, PcG proteins maintain genes in an "OFF" transcriptional state by stably modifying the chromatin configuration at gene loci. Two distinct PcG complexes have been characterized in *Drosophila*. The first complex, referred to as the Polycomb Repressive Complex 1 (PRC-1), is 1-2 MDa in size and is composed of PcG proteins Polycomb (PC), Polyhomeotic (PH), Posterior Sex Combs (PSC), dRING1, in addition to components of the general transcriptional machinery.[46-48] A role for PRC-1 in repressing transcription was elucidated by showing that PSC inhibits chromatin remodeling by preventing ATP-dependent chromatin remodeling complexes such as SWI/SNF from binding to a nucleosome array in vitro.[49] Although PRC-1 is conserved in mammals, mammalian PRC-1 does not seem to contain general transcription factors.[50] The second complex is Extra Sex Combs-Enhancer-of-Zeste or ESC-E(Z) and contains ESC, E(Z), the suppressor of position effect variegation SU(var)12, and the DNA-binding protein Pleihomeotic (PHO) - the homolog of mammalian YY1.[51-56] ESC-E(Z) is a histone methyl transferase (HMTase) whose function is required for proper PcG transcriptional repression, as well as for facilitating the binding of PRC-1 to its DNA targets.[53] Despite the biochemical distinction that separates PRC-1 and ESC-E(Z) complexes, these complexes appear to work together to repress *Hox* gene expression in vivo. Chromatin binding and repression by PC and PSC is dependent on the availability of E(Z).[57] Furthermore, direct interaction between ESC-E(Z) and PRC-1 components occurs in extracts obtained from early *Drosophila* embryos.[58] A model for the action of PcG complexes on chromatin templates of repressed genes would be as follows: First, PcG proteins belonging to the ESC-E(Z) complex are recruited to regulatory regions where they deacetylate and methylate N-terminal histone tails. This creates a histone code that results in further recruitment of members of the PRC-1 complex which maintains the closed chromatin state and blocks access to chromatin remodeling complexes.[59]

Several mammalian PcG homologs have been identified so far, and members of the mammalian PRC-1 complex include MEL18, BMI1, M33, and RING1.[60-66] Mice deficient in PcG proteins, including the three just mentioned, display anterior shifts in the expression boundaries of *Hox* genes in both neural and paraxial mesoderm tissue, showing particularly posterior transformations in the paraxial skeleton, consistent with ectopic *Hox* gene expression.[67-70] For example, in *mel18* mutant mice, both *Hoxb4* and *Hoxd4* expression are anteriorized and ectopically expressed in C1, resulting in an ectopic odontoid process normally characteristic of C2.[67] *Hoxd4* expression is also anteriorized in *mel18/bmi1* double mutants.[70] More importantly, anterior expression boundaries of several *Hox* genes in PcG mutant mice are normal at earlier developmental stages (E8.5), indicating that the functions of MEL18 and BMI1 are required to maintain, but not to initiate, the correct boundaries of *Hox* gene expression. A summary of PcG in vivo mutations and their resulting alterations in Hox expression and homeotic transformations in the mouse is listed in Table 1.

Results from a yeast two-hybrid screen identified a novel interaction between mammalian PcG proteins (Zfp144 and Rnf2) and Sf3b1, an essential spliceosomal protein.[71] Sf3b1 heterozygote mutant mice have decreased levels of Sf3b1 in PcG complexes and exhibit skeletal abnormalities as well as anterior shifts of expression of *Hoxb6*, *Hoxc6* and *Hoxb8* in prevertebrae. This suggests that the interaction between PcG proteins and members of the spliceosome machinery might be essential for PcG-mediated repression of Hox genes. Another study using mouse embryonic tissue showed that histone methylation at lysine 27 is required for the PcG protein Rnf2 to bind and repress the expression of *Hoxb8* in vivo.[72] Interestingly, some PcG mutations result in downregulation of *Hox* gene activity in their normal expression domain. Mutations in PcG genes *rnf10* and *Phc1* downregulate the expression of *Hoxb1*,[73] and decreased *Hoxb8* expression is observed in *Bmi1/Rnf10* and *Phc1/Phc2* double mutants.[70,71,73] The difference in *Hoxb8* transcriptional levels in these mutant mice is correlated with decreased histone H3 acetylation at K9, a marker of open chromatin.[72] These results indicate that PRC-1 complexes lacking RNF2 are involved in maintaining H3 acetylation in Hox-expressing embryonic tissue. This suggests that recruitment of PcG proteins to Hox genes might be required not only for maintenance of gene silencing, but also for maintaining transcriptional activation, also through modifying chromatin status.

Other studies have implicated the cell cycle regulator Geminin in regulating *Hox* gene expression. Geminin controls replication by binding to the licensing factor CDT1.[74] Geminin also associates with PcG members RAE2 and MEL18, and directly modulates the anterior expression boundary of *Hoxb9*, consistent with a PcG-like activity for this protein.[75] This raises the intriguing possibility that alterations to chromatin structure could be mechanistically linked to the cycle through Geminin function.

Table 1. In vivo PcG mutations and affected Hox genes in the mouse

PcG member	Affected *Hox* genes	References
Bmi1	a4, c4, a5, c5, b6, c6, c8, c9	68
Mel18	b3, b4, d4, a5, b6	67
Bmi1/Mel18	b3, d4, b6, c6, d13	70
M33	a3	90
Mph1	a3, b3, a4,b4, d4, d5	64
Rnf2	b4, b6, b8	69
Bmi1/M33	c9, c8	91

Trx Group Proteins

Four different TrxG complexes have been described in *Drosophila*: BRM, ASH1, ASH2, and TRX or TAC1,[76,77] all of which contain a SET domain which makes them likely candidates for chromatin remodeling by histone methylation.[12,78,79] The BRM complex is highly related to SWI/SNF and acts as an ATP-dependent chromatin remodeling complex.[80] The purification of TAC1 identified the HAT coactivator CBP as a member of *Drosophila* TAC1 complex and showed that it is essential for activating expression of the insect Hox gene *Ultrabithorax*.[77]

The human mixed-lineage leukemia gene (*MLL*) and mouse *mll* are highly homologous to *Drosophila Trx*, and naturally-occuring mutations of *MLL* result in acute lymphoid and myeloid leukemias. MLL-CBP fusions have also been identified in several human leukemias.[81,82] MLL-induced leukemias are probably the direct result of misregulation of *Hox* gene expression,[83] supported by the finding that MLL directly binds the *Hoxc8* proximal promoter resulting in transcriptional activation.[84] Moreover, reintroduction of MLL into $mll^{-/-}$ mouse embryonic fibroblast cells results in H3 methylation at K4 at Hox *loci* and the activation of *Hoxc8* transcription,[84] consistent with the demonstration that a stable complex forms between MLL1 and the H4 acetyltransferase MOF. Both proteins are crucial for *Hoxa9* transcriptional initiation, revealing roles for TrxG members beyond maintenance.[85]

Targeted disruption of *mll* results in smaller sized mice that display homeotic transformations in cervical, thoracic and lumbar vertebrae.[86] Both *Hoxa7* and *Hoxa9* anterior expression boundaries are posteriorized in the paraxial mesoderm of $mll^{-/-}$ mice at E9.5, and are no longer detectable at E10.5, emphasizing the importance of TrxG proteins in maintaining the transcriptional activation of *Hox* genes in the developing embryo. Further clarifying the mechanism of action of TrxG proteins, MLL interacts with RNA Pol II at active *Hox* loci up until hematopoiesis, after which *Hox* gene expression is sharply downregulated.[87] The loss of *mll* gene expression results in defects in RNA Pol II distribution. In addition, the targeted disruption of sequences encoding MLL's SET domain gives rise to skeletal abnormalities in mutant mice, and decreases the expression of several *Hox* genes including *Hoxd4*, *Hoxa7*, *Hoxa5* and *Hoxc8*[88] and altered DNA methylation patterns at the *Hoxd4* locus.

MOZ is a histone acetyltransferase of the MYST family required for expression of Hox paralogs 1 through 4 in the zebrafish hindbrain.[89] Loss of MOZ function does not affect the initiation of Hox gene expression, but rather its maintenance. Additionally, Hox expression and the *moz* mutant phenotype are rescued by the HDAC inhibitor trichostatin A, as is also the case in *Trx* mutants of *Drosophila*. Though not definitively proven, these findings have implicated MOZ in TrxG function.[89]

Conclusions

As for other transcription units, the regulation of Hox gene expression is elaborated, in part, at the level of chromatin. The organization of Hox genes into clusters, and the striking collinearity of their expression in time and space, demand the integration of cluster-wide and gene-specific mechanisms for controlling the status of chromatin. However complicated this may be in the embryonic trunk, it promises to be much more so in the tetrapod limb bud where HoxD genes are expressed in strikingly different patterns at early vs late times of limb morphogenesis, and come under the control of multiple enhancers located within and at great distances to either side of the cluster.[13,25] A mechanistic description of chromatin events during this developmental program is a major challenge.

References

1. McGinnis W, Hart CP, Gehring WJ et al. Molecular cloning and chromosome mapping of a mouse DNA sequence homologous to homeotic genes of Drosophila. Cell 1984; 38(3):675-680.
2. McGinnis W, Levine MS, Hafen E et al. A conserved DNA sequence in homoeotic genes of the Drosophila Antennapedia and bithorax complexes. Nature 1984; 308(5958):428-433.
3. Scott MP, Weiner AJ. Structural relationships among genes that control development: Sequence homology between the Antennapedia, Ultrabithorax, and fushi tarazu loci of Drosophila. Proc Natl Acad Sci USA 1984; 81(13):4115-4119.
4. Krumlauf R. Hox genes in vertebrate development. Cell 1994; 78(2):191-201.
5. Kappen C, Schughart K, Ruddle FH. Organization and expression of homeobox genes in mouse and man. Ann NY Acad Sci 1989; 567:243-252.
6. Manzanares M, Wada H, Itasaki N et al. Conservation and elaboration of Hox gene regulation during evolution of the vertebrate head. Nature 2000; 408(6814):854-857.
7. de la Serna IL, Ohkawa Y, Imbalzano AN. Chromatin remodelling in mammalian differentiation: Lessons from ATP-dependent remodellers. Nat Rev Genet 2006; 7(6):461-473.
8. Featherstone M. Coactivators in transcription initiation: Here are your orders. Curr Opin Genet Dev 2002; 12(2):149-155.
9. Fischle W, Wang Y, Allis CD. Histone and chromatin cross-talk. Curr Opin Cell Biol 2003; 15(2):172-183.
10. Roth SY, Denu JM, Allis CD. Histone acetyltransferases. Annu Rev Biochem 2001; 70:81-120.
11. Yang XJ, Seto E. Collaborative spirit of histone deacetylases in regulating chromatin structure and gene expression. Curr Opin Genet Dev 2003; 13(2):143-153.
12. Jenuwein T, Allis CD. Translating the histone code. Science 2001; 293(5532):1074-1080.
13. Kmita M, Duboule D. Organizing axes in time and space; 25 years of colinear tinkering. Science 2003; 301(5631):331-333.
14. Lufkin T, Mark M, Hart CP et al. Homeotic transformation of the occipital bones of the skull by ectopic expression of a homeobox gene. Nature 1992; 359(6398):835-841.
15. Duboule D. Vertebrate hox gene regulation: Clustering and/or colinearity? Curr Opin Genet Dev 1998; 8(5):514-518.
16. Duboule D, Deschamps J. Colinearity loops out. Dev Cell 2004; 6(6):738-740.
17. Roelen BA, de Graaff W, Forlani S et al. Hox cluster polarity in early transcriptional availability: A high order regulatory level of clustered Hox genes in the mouse. Mech Dev 2002; 119(1):81-90.
18. Deschamps J, van Nes J. Developmental regulation of the Hox genes during axial morphogenesis in the mouse. Development 2005; 132(13):2931-2942.
19. Rastegar M, Kobrossy L, Kovacs EN et al. Sequential histone modifications at Hoxd4 regulatory regions distinguish anterior from posterior embryonic compartments. Mol Cell Biol 2004; 24(18):8090-8103.
20. Bickmore WA, Mahy NL, Chambeyron S. Do higher-order chromatin structure and nuclear reorganization play a role in regulating Hox gene expression during development? Cold Spring Harb Symp Quant Biol 2004; 69:251-257.
21. Chambeyron S, Bickmore WA. Chromatin decondensation and nuclear reorganization of the HoxB locus upon induction of transcription. Genes Dev 2004; 18(10):1119-1130.
22. Cirillo LA, Lin FR, Cuesta I et al. Opening of compacted chromatin by early developmental transcription factors HNF3 (FoxA) and GATA-4. Mol Cell 2002; 9(2):279-289.
23. Berkes CA, Bergstrom DA, Penn BH et al. Pbx marks genes for activation by MyoD indicating a role for a homeodomain protein in establishing myogenic potential. Mol Cell 2004; 14(4):465-477.
24. Weston AD, Blumberg B, Underhill TM. Active repression by unliganded retinoid receptors in development: Less is sometimes more. J Cell Biol 2003; 161(2):223-228.
25. Tarchini B, Duboule D. Control of Hoxd genes' collinearity during early limb development. Dev Cell 2006; 10(1):93-103.
26. Weston AD, Hoffman LM, Underhill TM. Revisiting the role of retinoid signaling in skeletal development. Birth Defects Res C Embryo Today 2003; 69(2):156-173.
27. Dupé V, Davenne M, Brocard J et al. In vivo functional analysis of the Hoxa-1 3' retinoic acid response element (3'RARE). Development 1997; 124(2):399-410.
28. Gould A, Itasaki N, Krumlauf R. Initiation of rhombomeric Hoxb4 expression requires induction by somites and a retinoid pathway. Neuron 1998; 21(1):39-51.
29. Huang D, Chen SW, Gudas LJ. Analysis of two distinct retinoic acid response elements in the homeobox gene Hoxb1 in transgenic mice. Dev Dyn 2002; 223(3):353-370.
30. Langston AW, Gudas LJ. Identification of a retinoic acid responsive enhancer 3' of the murine homeobox gene Hox-1.6. Mech Dev 1992; 38(3):217-227.

31. Marshall H, Studer M, Pöpperl H et al. A conserved retinoic acid response element required for early expression of the homeobox gene Hoxb-1. Nature 1994; 370(6490):567-571.
32. Packer AI, Crotty DA, Elwell VA et al. Expression of the murine Hoxa4 gene requires both autoregulation and a conserved retinoic acid response element. Development 1998; 125(11):1991-1998.
33. Studer M, Pöpperl H, Marshall H et al. Role of a conserved retinoic acid response element in rhombomere restriction of Hoxb-1. Science 1994; 265(5179):1728-1732.
34. Zhang F, Nagy Kovacs E, Featherstone MS. Murine hoxd4 expression in the CNS requires multiple elements including a retinoic acid response element. Mech Dev 2000; 96(1):79-89.
35. Oosterveen T, Niederreither K, Dolle P et al. Retinoids regulate the anterior expression boundaries of 5' Hoxb genes in posterior hindbrain. EMBO J 2003; 22(2):262-269.
36. van der Hoeven F, Zákány J, Duboule D. Gene transpositions in the HoxD complex reveal a hierarchy of regulatory controls. Cell 1996; 85(7):1025-1035.
37. Kondo T, Duboule D. Breaking colinearity in the mouse HoxD complex. Cell 1999; 97(3):407-417.
38. Barna M, Merghoub T, Costoya JA et al. Plzf mediates transcriptional repression of HoxD gene expression through chromatin remodeling. Dev Cell 2002; 3(4):499-510.
39. Chambeyron S, Da Silva NR, Lawson KA et al. Nuclear reorganisation of the Hoxb complex during mouse embryonic development. Development 2005; 132(9):2215-2223.
40. Kobrossy L, Rastegar M, Featherstone M. Interplay between chromatin and trans-acting factors regulating the Hoxd4 promoter during neural differentiation. J Biol Chem 2006.
41. Narlikar GJ, Fan HY, Kingston RE. Cooperation between complexes that regulate chromatin structure and transcription. Cell 2002; 108(4):475-487.
42. Meccia E, Bottero L, Felicetti F et al. HOXB7 expression is regulated by the transcription factors NF-Y, YY1, Sp1 and USF-1. Biochim Biophys Acta 2003; 1626(1-3):1-9.
43. Gilthorpe J, Vandromme M, Brend T et al. Spatially specific expression of Hoxb4 is dependent on the ubiquitous transcription factor NFY. Development 2002; 129(16):3887-3899.
44. Bernstein BE, Kamal M, Lindblad-Toh K et al. Genomic maps and comparative analysis of histone modifications in human and mouse. Cell 2005; 120(2):169-181.
45. Kennison JA. The Polycomb and trithorax group proteins of Drosophila: Trans-regulators of homeotic gene function. Annu Rev Genet 1995; 29:289-303.
46. Shao Z, Raible F, Mollaaghababa R et al. Stabilization of chromatin structure by PRC1, a Polycomb complex. Cell 1999; 98(1):37-46.
47. Saurin AJ, Shao Z, Erdjument-Bromage H et al. A Drosophila Polycomb group complex includes Zeste and dTAFII proteins. Nature 2001; 412(6847):655-660.
48. Satijn DP, Otte AP. Polycomb group protein complexes: Do different complexes regulate distinct target genes? Biochim Biophys Acta 1999; 1447(1):1-16.
49. Francis NJ, Saurin AJ, Shao Z et al. Reconstitution of a functional core polycomb repressive complex. Mol Cell 2001; 8(3):545-556.
50. Levine SS, Weiss A, Erdjument-Bromage H et al. The core of the polycomb repressive complex is compositionally and functionally conserved in flies and humans. Mol Cell Biol 2002; 22(17):6070-6078.
51. Tie F, Furuyama T, Prasad-Sinha J et al. The Drosophila Polycomb Group proteins ESC and E(Z) are present in a complex containing the histone-binding protein p55 and the histone deacetylase RPD3. Development 2001; 128(2):275-286.
52. Ng J, Hart CM, Morgan K et al. A Drosophila ESC-E(Z) protein complex is distinct from other polycomb group complexes and contains covalently modified ESC. Mol Cell Biol 2000; 20(9):3069-3078.
53. Cao R, Wang L, Wang H et al. Role of histone H3 lysine 27 methylation in Polycomb-group silencing. Science 2002; 298(5595):1039-1043.
54. Czermin B, Melfi R, McCabe D et al. Drosophila enhancer of Zeste/ESC complexes have a histone H3 methyltransferase activity that marks chromosomal Polycomb sites. Cell 2002; 111(2):185-196.
55. Kuzmichev A, Nishioka K, Erdjument-Bromage H et al. Histone methyltransferase activity associated with a human multiprotein complex containing the Enhancer of Zeste protein. Genes Dev 2002; 16(22):2893-2905.
56. Muller J, Hart CM, Francis NJ et al. Histone methyltransferase activity of a Drosophila Polycomb group repressor complex. Cell 2002; 111(2):197-208.
57. Rastelli L, Chan CS, Pirrotta V. Related chromosome binding sites for zeste, suppressors of zeste and Polycomb group proteins in Drosophila and their dependence on Enhancer of zeste function. EMBO J 1993; 12(4):1513-1522.
58. Poux S, McCabe D, Pirrotta V. Recruitment of components of Polycomb Group chromatin complexes in Drosophila. Development 2001; 128(1):75-85.

59. Simon JA, Tamkun JW. Programming off and on states in chromatin: Mechanisms of Polycomb and trithorax group complexes. Curr Opin Genet Dev 2002; 12(2):210-218.
60. Tagawa M, Sakamoto T, Shigemoto K et al. Expression of novel DNA-binding protein with zinc finger structure in various tumor cells. J Biol Chem 1990; 265(32):20021-20026.
61. van Lohuizen M, Frasch M, Wientjens E et al. Sequence similarity between the mammalian bmi-1 proto-oncogene and the Drosophila regulatory genes Psc and Su(z)2. Nature 1991; 353(6342):353-355.
62. Pearce JJ, Singh PB, Gaunt SJ. The mouse has a Polycomb-like chromobox gene. Development 1992; 114(4):921-929.
63. Satijn DP, Olson DJ, van der Vlag J et al. Interference with the expression of a novel human polycomb protein, hPc2, results in cellular transformation and apoptosis. Mol Cell Biol 1997; 17(10):6076-6086.
64. Gunster MJ, Satijn DP, Hamer KM et al. Identification and characterization of interactions between the vertebrate polycomb-group protein BMI1 and human homologs of polyhomeotic. Mol Cell Biol 1997; 17(4):2326-2335.
65. Hemenway CS, Halligan BW, Levy LS. The Bmi-1 oncoprotein interacts with dinG and MPh2: The role of RING finger domains. Oncogene 1998; 16(19):2541-2547.
66. Schoorlemmer J, Marcos-Gutierrez C, Were F et al. Ring1A is a transcriptional repressor that interacts with the Polycomb-M33 protein and is expressed at rhombomere boundaries in the mouse hindbrain. EMBO J 1997; 16(19):5930-5942.
67. Akasaka T, Kanno M, Balling R et al. A role for mel-18, a Polycomb group-related vertebrate gene, during theanteroposterior specification of the axial skeleton. Development 1996; 122(5):1513-1522.
68. van der Lugt NM, Alkema M, Berns A et al. The Polycomb-group homolog Bmi-1 is a regulator of murine Hox gene expression. Mech Dev 1996; 58(1-2):153-164.
69. Suzuki M, Mizutani-Koseki Y, Fujimura Y et al. Involvement of the Polycomb-group gene Ring1B in the specification of the anterior-posterior axis in mice. Development 2002; 129(18):4171-4183.
70. Akasaka T, van Lohuizen M, van der Lugt N et al. Mice doubly deficient for the Polycomb Group genes Mel18 and Bmi1 reveal synergy and requirement for maintenance but not initiation of Hox gene expression. Development 2001; 128(9):1587-1597.
71. Isono K, Fujimura Y, Shinga J et al. Mammalian polyhomeotic homologues Phc2 and Phc1 act in synergy to mediate polycomb repression of Hox genes. Mol Cell Biol 2005; 25(15):6694-6706.
72. Fujimura Y, Koseki H. Role of mammalian polycomb group gene products in embryo genesis. Tanpakushitsu Kakusan Koso 2005; 50(6 Suppl):563-568.
73. de Graaff W, Tomotsune D, Oosterveen T et al. Randomly inserted and targeted Hox/reporter fusions transcriptionally silenced in Polycomb mutants. Proc Natl Acad Sci USA 2003; 100(23):13362-13367.
74. Luo L, Kessel M. Geminin coordinates cell cycle and developmental control. Cell Cycle 2004; 3(6):711-714.
75. Luo L, Yang X, Takihara Y et al. The cell-cycle regulator geminin inhibits Hox function through direct and polycomb-mediated interactions. Nature 2004; 427(6976):749-753.
76. Papoulas O, Beek SJ, Moseley SL et al. The Drosophila trithorax group proteins BRM, ASH1 and ASH2 are subunits of distinct protein complexes. Development 1998; 125(20):3955-3966.
77. Petruk S, Sedkov Y, Smith S et al. Trithorax and dCBP acting in a complex to maintain expression of a homeotic gene. Science 2001; 294(5545):1331-1334.
78. Lachner M, O'Carroll D, Rea S et al. Methylation of histone H3 lysine 9 creates a binding site for HP1 proteins. Nature 2001; 410(6824):116-120.
79. Zhang Y, Reinberg D. Transcription regulation by histone methylation: Interplay between different covalent modifications of the core histone tails. Genes Dev 2001; 15(18):2343-2360.
80. Kal AJ, Mahmoudi T, Zak NB et al. The Drosophila brahma complex is an essential coactivator for the trithorax group protein zeste. Genes Dev 2000; 14(9):1058-1071.
81. Sobulo OM, Borrow J, Tomek R et al. MLL is fused to CBP, a histone acetyltransferase, in therapy-related acute myeloid leukemia with a t(11; 16)(q23; p13.3). Proc Natl Acad Sci USA 1997; 94(16):8732-8737.
82. Lavau C, Du C, Thirman M et al. Chromatin-related properties of CBP fused to MLL generate a myelodysplastic-like syndrome that evolves into myeloid leukemia. EMBO J 2000; 19(17):4655-4664.
83. Hess JL. Mechanisms of transformation by MLL. Crit Rev Eukaryot Gene Expr 2004; 14(4):235-254.
84. Milne TA, Briggs SD, Brock HW et al. MLL targets SET domain methyltransferase activity to Hox gene promoters. Mol Cell 2002; 10(5):1107-1117.
85. Dou Y, Milne TA, Tackett AJ et al. Physical association and coordinate function of the H3 K4 methyltransferase MLL1 and the H4 K16 acetyltransferase MOF. Cell 2005; 121(6):873-885.

86. Yu BD, Hess JL, Horning SE et al. Altered Hox expression and segmental identity in Mll-mutant mice. Nature 1995; 378(6556):505-508.
87. Milne TA, Dou Y, Martin ME et al. MLL associates specifically with a subset of transcriptionally active target genes. Proc Natl Acad Sci USA 2005; 102(41):14765-14770.
88. Terranova R, Agherbi H, Boned A et al. Histone and DNA methylation defects at Hox genes in mice expressing a SET domain-truncated form of Mll. Proc Natl Acad Sci USA 2006; 103(17):6629-6634.
89. Marks PA, Richon VM, Miller T et al. Histone deacetylase inhibitors. Adv Cancer Res 2004; 91:137-168.
90. Core N, Bel S, Gaunt SJ et al. Altered cellular proliferation and mesoderm patterning in Polycomb-M33-deficient mice. Development 1997; 124(3):721-729.
91. Bel S, Core N, Djabali M et al. Genetic interactions and dosage effects of Polycomb group genes in mice. Development 1998; 125(18):3543-3551.

Chapter 7

Homeobox-Containing Genes in Limb Regeneration

David M. Gardiner* and Susan V. Bryant

Abstract

Early investigations established an important role for homeobox-containing genes in the initiation of regeneration, as well as in the later pattern formation events leading to a new limb. The recent increased research on the mechanisms of regeneration, along with the fact that urodele amphibians provide the only opportunity to understand how vertebrates can regenerate their limbs, has led to a renewed interest in the functioning of this important group of genes during salamander limb regeneration. It appears that all vertebrates, including humans, have impressive regenerative abilities as embryos; however, all but urodeles lose much of these abilities as development proceeds. In contrast, cells in adult urodeles are unique in their ability to revert to an embryonic state (dedifferentiate) in order to recapitulate embryogenesis, Consequently, urodeles are the only adult vertebrates that can completely and perfectly regenerate entire limbs, and thus they offer a unique opportunity to gain critical insights for future advances in regenerative medicine. Much data indicate that a large number of homeobox genes play important roles in the initiation and regulation of limb regeneration. In some instances, the regulatory mechanisms controlling homeobox gene expression appear comparable to what is observed in developing limbs; whereas, in others they different dramatically. In spite of differences in spatial and temporal expression patterns, homeobox gene function is conserved in both regeneration and development. Research on the role of homeobox genes is poised to move forward, particularly in the context of the early stages that are unique to regeneration, and thus are critical in achieving the goal of inducing human regeneration. These efforts will be possible because of the new genetic resources for research utilizing the axolotl as a model system.

Homeobox-containing genes were among the first genes identified as having a significant function in the regulation of embryonic development. Although the pioneering work was carried out with *Drosophila*, it soon became apparent that the structure and function of these genes is highly conserved, and that they play important roles in vertebrate development. Particularly evident was their function in the control of body and appendage pattern, thus validating the premolecular biology predictions that the mechanisms controlling pattern formation would be conserved among such divergent organisms as flies, grasshoppers and salamanders.[1-3] It thus was not long before studies began to demonstrate a role for homeobox genes in the control of salamander limb regeneration.

After an exciting start, investigations into the role of homeobox genes in regeneration has languished in recent years. Fortunately, there is currently a much increased interest in

*Corresponding Author: David M. Gardiner—Department of Developmental and Cell Biology, 4111 Natural Sciences II, University of California, Irvine, California 92697-2275, U.S.A.
Email: dmgardin@uci.edu

HOX Gene Expression, edited by Spyros Papageorgiou. ©2007 Landes Bioscience and Springer Science+Business Media.

regeneration, along with a renewed appreciation of the fact that urodele amphibians provide the only opportunity to understand how vertebrates can regenerate their limbs. It appears that all vertebrates, including humans, have impressive regenerative abilities as embryos; however, all but urodeles lose much of these abilities as development proceeds (see ref. 4). In contrast, cells in adult urodeles are unique in their ability to revert to an embryonic state (dedifferentiate) in order to recapitulate embryogenesis (see ref. 5). Consequently, urodeles are the only adult vertebrates that can completely and perfectly regenerate entire limbs, and thus they offer a unique opportunity to gain critical insights for future advances in regenerative medicine.

Data indicate that several homeobox genes play important roles in the initiation and regulation of limb regeneration. Our goal in writing this review is to stimulate future efforts in the field of limb regeneration research. We have elected to not consider data on the regeneration of limb buds, because they do not provide insights into the early critical events of dedifferentiation and blastema formation that are unique to adult urodeles. It is these early steps that will need to be induced in order to stimulate human regeneration. The cells of the embryo by contrast, are already immature, thus bypassing the need for dedifferentiation. Nevertheless, there is evidence for important functional roles for homeobox genes in limb bud regeneration (e.g., see refs. 6,7). In this chapter, we begin by reviewing the early studies that established the importance of homeobox genes in adult limb regeneration. In the second half, we focus on what we consider to be the emerging areas of limb regeneration research with respect to the function of homeobox genes.

Early Studies

The earliest studies identified Hox genes expressed during limb regeneration in the newt (see ref. 8). As each was identified and studied, it was noted that they often were also expressed in the normal, unamputated limb tissues, leading to the proposal that sustained Hox expression might be causally related to the regenerative ability of the adult urodele limb.[9,10] With subsequent studies, it appears that is not the case for most of the homeobox genes (see ref. 8), and the issue of the significance of Hox expression in adult limbs is as yet unresolved. It is relevant to note that fingers and toes of adult salamanders are easily injured, and thus may be in a state of nearly continuous regeneration. This phenomenon has been observed for expression of blastema markers in unamputated finger tips.[11] Thus the inclusion of distal structures in RNA preparations would give the impression of sustained expression in the adult limb.[8]

A large scale screen for homeobox genes expressed during axolotl limb regeneration, made it apparent that there is not a single homeobox gene, or even just a few, but many that are involved in the regulation of growth and pattern formation during limb regeneration.[12] To date, about 24 homeobox genes have been reported to be expressed in regenerating urodele limbs (see refs. 8,12). Over time, it was also observed that some of these genes are expressed as multiple transcripts with spatially distinct expression patterns[13,14] (Gardiner, unpublished data), indicating an even more complex role in regeneration that originally envisioned.

With the optimization of techniques for in situ hybridization to urodele tissues,[12] it became possible to analyze the spatial and temporal details of homeobox expression. Such analyses led to a new appreciation of the complexity of the early events of regeneration. Almost all previous studies had focused on the behavior of cells relatively late in regeneration, after the blastema had formed. From a number of grafting and cell lineage studies it appeared that limb development (limb buds) and limb regeneration (blastemas) used the same mechanisms for the later stages of control of growth and pattern formation.[15] Very little was known about mechanisms controlling events prior to blastema formation. The early studies of expression of the HoxA complex genes (see below) led to the realization that there are at least two distinct phases to regeneration. The first, early phase begins with wound healing and ends with the formation of a regeneration blastema. This is followed by the second, redevelopment phase that is a recapitulation of the events that occurred during limb development in the embryo.[5,16] Although the later phase of limb regeneration is equivalent to limb development, the early phase that results in genesis of the blastema is unique to regeneration. Thus the unique events of regeneration involve the rapid closure of the wound by epidermal cells, interactions between the wound epidermis and underlying stump cells that stimulate dedifferentiation, and cell migration and proliferation that give rise to the blastema. Dedifferentiation of stump cells to give rise to the relatively undifferentiated blastema cells, though still not well understood mechanistically, is clearly a requisite step in the transformation from the amputated mature limb to the regenerating blastema.

Based on the previous studies indicating the similarities of limb development and regeneration, we had anticipated that the spatial and temporal patterns of HoxA expression in regeneration would be comparable to that observed during limb development. In contrast, expression differed dramatically in both time and space. The induction of *Hoxa-9* and *Hoxa-13* occurs surprisingly soon after injury, at least within 24 hours,[12] and is thus one of the earliest reported molecular events specific to regeneration. Expression of several other genes is induced earlier, but they are also expressed in lateral skin wounds, which do not form outgrowths or accessory limbs. Both *Hoxa-9* and *Hoxa-13* are expressed in the distal-most cells of the amputated stump, which several days later will give rise to the early blastema (Fig. 1). The early blastema subsequently increases in size as a consequence of continued recruitment of cells from the stump,

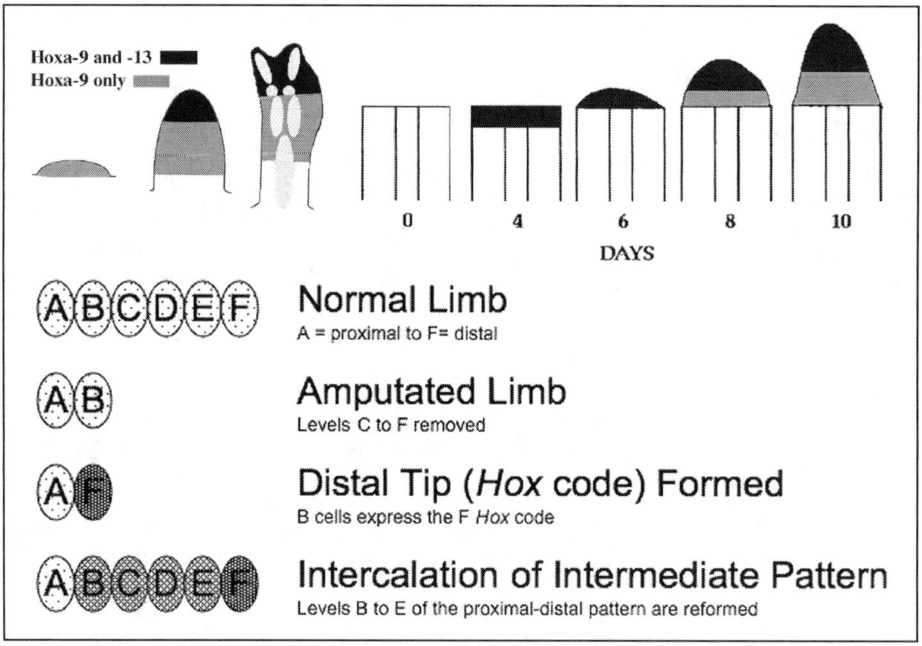

Figure 1. Establishment of the distal tip of the amputated limb, and the subsequent intercalation of the missing positional values between the distal tip and the stump. The temporal and spatial patterns of expression of the 5' HoxA genes (*Hoxa-9* and *Hoxa-13*) are the same as in other developing vertebrate limb buds (upper left). *Hoxa-9* is expressed at an earlier stage of limb development, and *Hoxa-13* has more distally restricted domain of expression. In contrast, during limb regeneration (upper right), these two genes are coexpressed in the distal cells of the amputate limb stump soon after amputation (Day 4) and in the early bud blastema (Day 6). With subsequent growth of the blastema, the expression domains become comparable to what is observed in developing limb buds, with *Hoxa-9* expressed through out the blastema and *Hoxa-13* expressed on in the distal tip. The sequence of events in the restoration of the amputated limb pattern is illustrated and annotated in the lower half of the figure in which positional information ranging from the most proximal (A value) to the most distal (F value) is indicated.

and cellular proliferation. As the blastema grows, a region of cells expressing *Hoxa-9* but not *Hoxa-13* is generated at the base of the blastema. When the regenerated skeletal elements begin to differentiate, both *Hoxa-9* and *Hoxa-13* are expressed in the autopod, whereas *Hoxa-9*, but not *Hoxa-13*, is expressed in the zeugopod. This final spatial expression pattern is the same as in developing limbs in urodeles as well as other vertebrates. Based on molecular and genetic evidence, specification of the distal-most region of the pattern (autopod) is a consequence of the coexpression of both 5' and more 3' members of the HoxA complex. The early coexpression *Hoxa-9* and *Hoxa-13* in stump cells indicates that regeneration is initiated by the reestablishment of the distal-most part of the limb pattern, regardless of the level of amputation. The more proximal regions of the pattern arise subsequently as a consequence of growth of the blastema and the intercalation of intermediate parts of the pattern (Fig. 1).[17] The early establishment of the distal tip of the limb ensures that the regenerated tissues will always be an exact replacement of the portion of the pattern that is removed.

Subsequent studies provided additional evidence of conserved function and expression of homeobox genes in both development and regeneration. This remarkable degree of conservation occurs in spite of differences in patterns of development. Urodeles differ from other tetrapods in that the distal most pattern (digits) differentiates in sequence from anterior to posterior, which is the reverse of what occurs in other vertebrates. This difference had previously been interpreted to reflect an underlying difference in the mechanisms of pattern formation.[18] An analysis of the pattern of HoxD expression revealed that this is not the case.[19] HoxD genes function in specification of the anterior-posterior pattern in the developing autopod, and exhibit the same pattern of expression in developing limb buds of the axolotl, in spite of the reverse order of differentiation. Similarly the more 5' HoxD genes are expressed in the more posterior cells of the regenerating limb, indicating a conserved function in specifying posterior positional information in both development and regeneration.

Although both the pattern of expression and function of homeobox genes are largely conserved in regenerating limbs as compared to developing limb buds, there is at least one notable exception. Genes within the HoxC complex are involved in specification of positional identity along the rostral-caudal axis of vertebrate embryos. *Hoxc-10* is expressed in developing hindlimbs and tails, but not in the forelimbs of either urodele larvae or of other vertebrate embryos.[14,20] *Hoxc-10* is however expressed at high levels in response to forelimb amputation in axolotls.[14] This is the only example to date of a regeneration-specific pattern of gene expression. In all other cases, genes expressed in regenerating limbs are also expressed during limb development. Thus *Hoxc-10* expression in regenerating forelimbs indicates the presence of regeneration-specific signals. Presumably there are elements in the promoter region of the axolotl *Hoxc-10* gene that are responsive to these signals, and provide an opportunity for the signals to be isolated and identified.

Two other models for studies of post-natal limb regeneration are the mouse digit-tip (discussed below) and the post-metamorphic *Xenopus* forelimb. In contrast to most nonurodele vertebrates that do not initiate a regenerative response, postmetamorphic *Xenopus* froglets regenerate a symmetrical cartilaginous spike in response to limb amputation. Studies comparing results from this model with the urodele limb regeneration model have provided insights about the distinction between regulatory pathways controlling outgrowth as compared to pattern formation,[6] and the distinction between nerve-dependent and –independent events in blastema formation.[21] As in urodeles, the HoxA genes are expressed, which is consistent with their function in regulating formation of the proximal-distal limb axis leading to limb outgrowth. *Msx1* is also expressed, which is consistent with its function in maintaining cells in a regeneration-competent state as demonstrated in the regenerating mouse digit tip (see below) and regenerating urodele limb.[22-24] In contrast, *shh* which is associated with regulation of anterior-posterior asymmetry, is not expressed in *Xenopus* regeneration, but is expressed in urodeles.[6,25,26] Regulation of the asymmetrically expressed HoxD genes, which are activated in urodele regeneration,[19] has not yet been investigated, but presumably they are not expressed in regenerating *Xenopus* froglet forelimbs.

Future Directions

The modern field of limb regeneration research is moving forward because the limb cells that are involved in the early events of regeneration leading to the formation of the blastema have been identified; at least three genetic pathways regulated by homeodomain-containing proteins have been identified as being involved in the critical initiation steps of limb regeneration; and, a new assay is available that will allow for the identification of the other critical signaling pathways controlling limb regeneration.

Blastema Forming Cells

The limb is composed of many cell types; however, it is the connective tissue fibroblasts that regulate regeneration both in terms of contribution to the blastema and the control of growth and pattern formation (see refs. 5,27). Though dermal fibroblasts account for less that 20% of all cells in the mature limb, they give rise to nearly 50% of the early/medium bud blastema cells on average, and as much as 78% of the cells at the maximum.[28] Since half of all limb fibroblasts are located in the dermis,[29] it is likely that the other half of the limb fibroblasts give rise to the other 50% of the blastema cells. In addition, fibroblast-containing tissues are the only limb tissues that influence growth and pattern formation during regeneration. Finally, entire limbs with normal pattern (though missing muscle) can be regenerated from fibroblasts as the sole source of progenitor cells (see ref. 5). In turn, these cells interact with the specialized wound epidermis derived from the basal keratinocytes of the mature skin epidermis leading to regeneration of the limb. Thus as the field moves forward, it will be important to focus on the role of homeobox genes in regulating the behavior of fibroblasts and basal keratinocytes as they undergo dedifferentiation allowing for the interactions controlling outgrowth and pattern formation.

Genetic Pathways

As discussed above, both the HoxA and HoxC genes have been identified as being involved in the critical initiation steps of limb regeneration. The very early, and initially overlapping expression of *Hoxa-9* and *Hoxa-13* indicates that one of the earliest steps in regeneration is the specification of the distal tip of the limb. It is likely that the interactions between the newly specified distal-tip cells and the more proximal stump cells are the stimulus leading to the intercalary growth that is known from classic studies to be the driving force of regenerative outgrowth.[1] Although there is much to be learned about the mechanisms of intercalary growth, they likely will also be critical in the induction of appendage regeneration in humans. Given that *Hoxc-10* is uniquely expressed in regenerating axolotl forelimbs, its regeneration-specific expression indicates the presence of regeneration-specific signals. Presumably there are elements in the promoter region of the axolotl *Hoxc-10* gene that are responsive to these signals, and thus provide an opportunity to isolate and identify those signals. Identification of these signals would then allow for the discovery of additional target pathways that are coordinately regulated during limb regeneration.

Most recently, the homeodomain-containing *Msx* genes have become the first genes with a demonstrated genetic function in regulating regeneration in a tetrapod limb.[30] Digit tips (distal to the last phalangeal joint) are able to regenerate both in embryonic and postnatal mice (see ref. 31). The domain of regeneration competence corresponds with *Msx1* expression. Digit tip regeneration is inhibited in the *Msx1* mutant, and regenerative failure can be rescued by BMP4.[30] *Msx* genes have also been implicated in the regulation of urodele limb regeneration.[22-24] It thus appears that the *Msx* genes have a conserved function in regenerative wound healing and dedifferentiation by repressing the differentiated phenotype of stump cells.[32]

The importance of *Msx* genes is also evident from studies of urodele muscle regeneration. Muscle progenitor cells appear to arise both from stem cells (satellite cells) as in other vertebrates,[33] and from fragmentation of myotubes to give rise to mononucleate myo-progenitor cells.[34,35] The relative contribution to regenerated muscle by these two mechanisms has not yet been determined. In addition, a multipotential murine cell line (C2C12) that can form a multi-nucleate, post-mitotic cell (myofiber) in vitro, can be induced to reenter the cell cycle, and fragment to give rise to mononucleate cells with the same multipotentiality of the parental cell line.[36,37] In both of these experimental models, *Msx1* mediates myotube dedifferentiation (fragmentation). In the case of C2C12 cells, myotube dedifferentiation can also be stimulated by an extract from regenerating newt limb blastemas.[36] Given that *Msx1* is already known to be involved in controlling this response, it is likely that the blastema-derived signals are operating upstream of the *Msx* transcription factors. Thus future studies of the function and regulation of *Msx* gene expression likely will identify signaling molecules expressed by regenerating limb cells that are involved in the control of dedifferentiation, growth and differentiation.

Regeneration Assay

Although the presence of a number of regeneration signals has been known for a long time, until recently most attempts to identify these signals have been unsuccessful. This lack of progress has been due largely to the lack of an appropriate assay to identify and test the function of such signals. Ironically, it is not possible to test the ability of a gene to induce a regenerative response in an amputated urodele limb since regeneration is the default response. Recently, a series of classical experimental manipulations to induce accessory limb formation have been modified to establish a novel assay (Accessory Limb Model, ALM) for identifying the signals that induce regeneration.[38] In this assay, a wound is created on an axolotl arm, which in the absence of any additional signals, heals without forming a scar. If however, a nerve is deviated to the wound, the local population of fibroblasts is induced to dedifferentiate and form an ectopic blastema. Finally, when a piece of skin is grafted contralaterally (anterior to posterior, or vice versa) to the site of a nerve-deviated wound, the ectopic blastemas are induced to form an ectopic limb. The ALM thus offers the advantage of being able to test candidate signals in a positive regeneration response assay, and allows for the identification of the signals that are unique to each of the events that occur during the successive phases of regeneration. The ALM demonstrates that critical signals between the wound epidermis, nerves, and fibroblasts are necessary and sufficient to induce dedifferentiation and the formation of a new limb. It also demonstrates how successful limb regeneration as a multi-step process, and that specific signals induce progression to the next step (Fig. 2).

The ALM has focused attention again on one of the earliest studied phenomena of limb regeneration, the role of nerves in allowing for successful regeneration. Although the importance of nerves in regeneration has long been recognized, the molecular mechanisms by which they influence the critical cellular interactions are largely unknown (see refs. 5,27). If the limb is denervated during the early stages of regeneration, it fails to regenerate. Thus, nerves are thought to produce a neurotrophic factor required for the initiation and progression of the early stages of regeneration. In the ALM, nerves provide a necessary signal(s) for the induction of an ectopic blastema and ultimately an accessory limb.[38] One target of the neurotrophic factor is the homeodomain-containing gene *Dlx3*.[39] *Dlx-3* is expressed at the distal region of regenerating limb blastemas, and based on its function in a number of other model systems, is likely to function in the specification of the distal tip of the limb. In regenerating axolotl limbs, *Dlx-3* expression is down regulated when a regenerating limb is denervated, and the limb fails to regenerate. However, delivery of exogenous FGF2 maintains *Dlx-3* expression and rescues regeneration. Thus the ALM provides the opportunity to identify the regulatory mechanisms for several important homeobox mediated pathways, including the HoxA, *Msx* and *Dlx* genes.

In conclusion, there has been a resurgence of research efforts on limb regeneration in recent years. In spite of the historically small size of the field, remarkable progress has been made recently in establishing the axolotl (*Ambystoma mexicanum*) as a model genetic organism. This progress has been facilitated by advances in genomics, bioinformatics, and somatic cell transgenesis in other fields, that have created the opportunity to investigate the mechanisms of important biological properties, such as limb regeneration. The *Ambystoma* Genetic Stock Center at the University of Kentucky is supported by the National Science Foundation as a resource

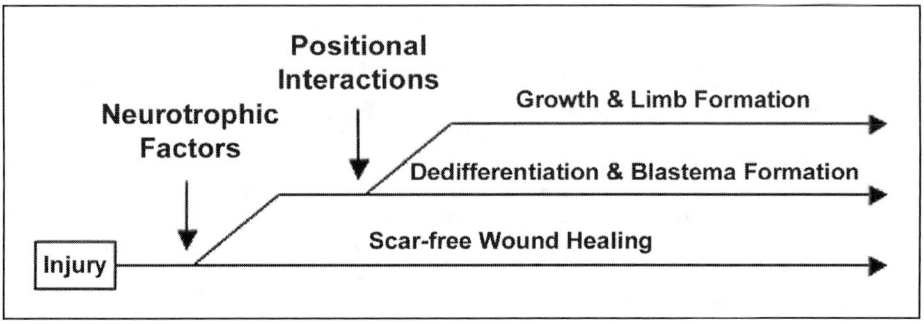

Figure 2. The stepwise sequence of events involved in the initiation of dedifferentiation and blastema formation. There are at least three critical events that lead to blastema formation and subsequent limb regeneration. The first is the creation of a wound and associated wound epidermis, that under the influence of nerve-derived factors give rise to a blastema. The presence of cells from different positions around the limb circumference leads to the positional interactions that induce the blastema to form a limb de novo.

for experimental work using the axolotl. Efforts supported by the National Center for Research Resources at the National Institutes of Health have led to the establishment of the Salamander Genome Project, which includes the creation of the first amphibian genetic linkage map[40] and several annotated molecular EST databases.[41-43] Thus the field of limb regeneration is at the point where significant discoveries from studies of animals that can regenerate (axolotl) will lead to insights for devising novel strategies to induce regeneration in animals that cannot (human). From what we already know, many of these discoveries, particularly with regards to the most critical early events of regeneration, will involve an understanding the role of homeobox genes, as discussed above. It is likely that the critical breakthroughs in regeneration research will come from understanding the mechanisms controlling expression of the homeobox genes and the function of their downstream targets.

Acknowledgements

Thanks to the many members of the Bryant/Gardiner Lab who have contributed to the ideas presented in this article, especially Dr. Tetsuya Endo (Hokkaido University) and Dr. Ken Muneoka (Tulane University, LA). Our research is supported by PHS grant HD-33465 and the National Science Foundation through its support of the *Ambystoma* Genetic Stock Center at the University of Kentucky, Lexington, KY.

References

1. Bryant SV, French V, Bryant PJ. Distal regeneration and symmetry. Science 1981; 212:993-1002.
2. French V, Bryant PJ, Bryant SV. Pattern regulation in epimorphic fields. Science 1976; 193:969-981.
3. Wolpert L. Positional information and pattern formation. Curr Top Dev Biol 1971; 6:183-224.
4. Gardiner DM. Ontogenetic decline of regenerative ability and the stimulation of human regeneration. Rejuvenation Res 2005; 8(3):141-153.
5. Bryant SV, Endo T, Gardiner DM. Vertebrate limb regeneration and the origin of limb stem cells. Int J Dev Biol 2002; 46(7):887-896.
6. Endo T, Tamura K, Ide H. Analysis of gene expressions during Xenopus forelimb regeneration. Dev Biol 2000; 220(2):296-306.
7. Hayamizu TF, Wanek N, Taylor G et al. Regeneration of HoxD expression domains during pattern regulation in chick wing buds. Dev Biol 1994; 161:504-512.
8. Geraudie J, Ferretti P. Gene expression during amphibian limb regeneration. Int Rev Cytol 1998; 180:1-50.
9. Savard P, Gates PB, Brockes JP. Position dependent expression of a homeobox gene transcript in relation to amphibian limb regeneration. EMBO J 1988; 7:4275-4282.

10. Simon HG, Tabin CJ. Analysis of Hox-4.5 and Hox-3.6 expression during newt limb regeneration: Differential regulation of paralagous Hox genes suggest different roles for members of different Hox clusters. Development 1993; 117:1397-1407.
11. Ferretti P, Ghosh S. Expression of regeneration-associated cytoskeletal proteins reveals differences and similarities between regenerating organs. Dev Dyn 1997; 210(3):288-304.
12. Gardiner DM, Blumberg B, Komine Y et al. Regulation of HoxA expression in developing and regenerating axolotl limbs. Development 1995; 121:1731-1741.
13. Beauchemin M, Savard P. Expression of five homeobox genes in the adult newt appendages and regeneration blastemas. In: Fallon JF et al, eds. Limb Development and Regeneration. New York: Wiley-Liss, 1993.
14. Carlson MRJ, Komine Y, Bryant SV et al. Expression of Hoxb13 and Hoxc10 in developing and regenerating axolotl limbs and tails. Dev Biol 2001; 229:396-406.
15. Muneoka K, Bryant SV. Evidence that patterning mechanisms in developing and regenerating limbs are the same. Nature 1982; 298:369-371.
16. Gardiner DM, Endo T, Bryant SV. The molecular basis of amphibian limb regeneration: Integrating the old with the new. Semin Cell Dev Biol 2002; 13(5):345-352.
17. Gardiner DM, Bryant SV. The tetrapod limb. In: Ferretti P, Geraudie J, eds. Cellular and Molecular Basis of Regeneration: From Invertebrates to Humans. New York: Wiley and Sons, Inc., 1998:187-205.
18. Shubin NH, Alberch P. A morphogenetic approach to the origin and basic organization of the tetrapod limb. Evol Biol 1986; 20:319-387.
19. Torok MA, Gardiner DM, Shubin NH et al. Expression of HoxD genes in developing and regenerating axolotl limbs. Dev Biol 1998; 200:225-233.
20. Peterson RL, Papenbrock T, Davda MM et al. The murine HoxC cluster contains five neighboring AbdB-related Hox genes that show unique spatially coordinated expression in posterior embryonic subregions. Mech Dev 1994; 47:253-260.
21. Suzuki M, Satoh A, Ide H et al. Nerve-dependent and -independent events in blastema formation during Xenopus froglet limb regeneration. Dev Biol 2005; 286(1):361-375.
22. Carlson MRJ, Bryant SV, Gardiner DM. Expression of Msx-2 during development, regeneration, and wound healing in axolotl limbs. J Exp Zool 1998; 282(6):715-723.
23. Koshiba K, Kuroiwa A, Yamamoto H et al. Expression of Msx genes in regenerating and developing limbs of axolotl. J Exp Zool 1998; 282(6):703-714.
24. Simon HG, Nelson C, Goff D et al. Differential expression of myogenic regulatory genes and Msx-1 during dedifferentiation and redifferentiation of regenerating amphibian limbs. Dev Dynamics 1995; 202:1-12.
25. Imokawa Y, Yoshizato K. Expression of Sonic hedgehog gene in regenerating newt limb blastema recapitulates that in developing limb buds. Proc Natl Acad Sci USA 1997; 94(17):9159-9164.
26. Torok MA, Gardiner DM, Izpisua-Belmonte JC et al. Sonic hedgehog (shh) expression in developing and regenerating axolotl limbs. J exp Zool 1999; 284:197-206.
27. Mescher AL. The cellular basis of limb regeneration in urodeles. Int J Dev Biol 1996; 40:785-795.
28. Muneoka K, Fox W, Bryant SV. Cellular contribution from dermis and cartilage to the regenerating limb blastema in axolotls. Dev Biol 1986; 116:256-260.
29. Tank PW, Holder N. The distribution of cells in the upper forelimbs of the axolotl. J Exp Zool 1979; 209:435-442.
30. Han M, Yang X, Farrington JE et al. Digit regeneration is regulated by Msx1 and BMP4 in fetal mice. Development 2003; 130(21):5123-5132.
31. Muller TL, Ngo-Muller V, Reginelli A et al. Regeneration in higher vertebrates: Limb buds and digit tips. Semin Cell Dev Biol 1999; 10:405-413.
32. Lee H, Habas R, Abate-Shen C. MSX1 cooperates with histone H1b for inhibition of transcription and myogenesis. Science 2004; 304(5677):1675-1678.
33. Cameron JA, Hilgers AR, Hinterberger TJ. Evidence that reserve cells are a source of regenerated adult newt muscle in-vitro. Nature 1986; 321(6070):607-610.
34. Kumar A, Velloso CP, Imokawa Y et al. Plasticity of retrovirus-labelled myotubes in the newt limb regeneration blastema. Dev Biol 2000; 218:125-136.
35. Lo DC, Allen F, Brockes JP. Reversal of muscle differentiation during urodele limb regeneration. Proc Natl Acad Sci USA 1993; 90(15):7230-7234.
36. McGann CJ, Odelberg SJ, Keating MT. Mammalian myotube dedifferentiation induced by newt regeneration extract. Proc Natl Acad Sci USA 2001; 98(24):13699-13704.
37. Odelberg SJ, Kollhoff A, Keating MT. Dedifferentiation of mammalian myotubes induced by msx1. Cell 2000; 103:1099-1109.

38. Endo T, Bryant SV, Gardiner DM. A stepwise model system for limb regeneration. Dev Biol 2004; 270(1):135-145.
39. Mullen L, Bryant SV, Torok MA et al. Nerve dependency of regeneration: The role of Distal-less and FGF signaling in amphibian limb regeneration. Development 1996; 122(11):3487-3497.
40. Smith JJ, Kump DK, Walker JA et al. A comprehensive expressed sequence tag linkage map for tiger salamander and Mexican axolotl: Enabling gene mapping and comparative genomics in Ambystoma. Genetics 2005; 171(3):1161-1171.
41. Habermann B, Bebin AG, Herklotz S et al. An Ambystoma mexicanum EST sequencing project: Analysis of 17,352 expressed sequence tags from embryonic and regenerating blastema cDNA libraries. Genome Biol 2004; 5(9):R67.
42. Putta S, Smith JJ, Walker JA et al. From biomedicine to natural history research: EST resources for ambystomatid salamanders. BMC Genomics 2004; 5(1):54.
43. Smith JJ, Putta S, Walker JA et al. Sal-Site: Integrating new and existing ambystomatid salamander research and informational resources. BMC Genomics 2005; 6:181.

CHAPTER 8

Hox Genes and Stem Cells

Mina Gouti and Anthony Gavalas*

Introduction

Stem cells are cells that undergo self-renewal as well as differentiation into progenitor cells. They are abundantly present, although ill defined, during development and it is believed that most, if not all, adult tissues harbor small populations of stem cells. Adult stem cells have been described for intestine, skin, muscle, blood and nervous system and may provide a tissue specific resource for tissue damage repair. A balance between stem cell self-renewal and differentiation maintains homeostasis in adult tissue. Thus normal tissue stem cells are defined by three common properties. (1) extensive self-renewal capacity, (2) strict control of stem cell numbers and (3) ability to undergo extensive differentiation to reconstitute all the functional elements in a given tissue. In this chapter we review the evidence that Hox genes may be involved in stem cell maintenance and control of self-renewal in different cellular and developmental contexts. Strong evidence exists for their role in controlling Drosophila neuroblast numbers and fate decisions and for their role in controlling expansion, self-renewal and lineage specification in the mammalian hematopoietic stem cells. There are tantalizing clues for their involvement in stem cell expansion during embryonic development and later in the process of oncogenesis.

Hox Genes and Drosophila Neuroblasts

Regulation of stem cell division is of particular interest for both developmental studies and stem cell therapeutics. The Drosophila central nervous system is generated by stem-cell-like post-embryonic progenitors called neuroblasts (pNBs). Drosophila neuroblasts are similar to mammalian neural stem cells in that they self-renew and have the potential to generate many different types of neurons and glia.[1] As such they are set to provide major insights concerning neural stem cell initiation, maintenance, termination and the mechanisms controlling asymmetric and symmetric stem cell division. As the number and identity of neurons produced at different segments of the developing larva differ, AP positional identity conferred by the Hox gene axial patterning system plays an important role in regulating these differences.

Such differences are at least in part due to differential rates of apoptosis[2,3] of the pNBs. Region specific apoptosis in pNBs is induced by a pulse of expression of the Hox gene AbdA which is differentially expressed along the AP axis of the larva.[4] In the absence of either *AbdA* or the proapoptotic genes *reaper*, *hid* and *grim* abdominal NBs are rescued from cell death and continue to proliferate. Bello et al showed that in pNBs the absence of postembryonic *AbdA* expression results in the expansion of their neuronal progeny, a result similar to the one induced by the loss of specific cell death genes. Furthermore, ectopic expression of *AbdA* in the thorax resulted in excessive cell death and significant subsequent reduction of neuronal clones. Therefore, *AbdA* regulates neuroblast elimination in the late embryo by inducing programmed

*Corresponding Author: Anthony Gavalas—Foundation for Biomedical Research of the Academy of Athens, Soranou Efessiou 4, Athens 11527, Hellas/Greece. Email: agavalas@bioacademy.gr

HOX Gene Expression, edited by Spyros Papageorgiou. ©2007 Landes Bioscience and Springer Science+Business Media.

cell death involving proapoptic genes and this provides an irreversible mechanism ensuring that embryonic neuroblasts stop dividing. Interestingly, ectopic expression of other Hox genes such as Ubx and Antp can also induce apoptosis of thoracic neuroblasts in a neuroblast cell autonomous manner.[4] This may be a general theme at final stages of differentiation as the Hox genes Deformed and Abdominal-B have been shown to shape segments in the developing Drosophila head by selectively activating reaper and thereby promoting programmed cell death.[5] The study of Hox genes in several systems has led to the concept of cellular memory. Once a characteristic set of Hox genes is activated its expression is maintained in the progeny of that lineage. Bello et al have shown that in the context of pNB it is imperative that such expression is abrogated so that in subsequent developmental stages a pulse of expression can shape the final size of the neuronal progeny. This is a markedly different mode of operation and the question arises whether other Hox genes are similarly regulated. This is particularly intriguing since heterotopic transplantation experiments have suggested that pNB lineages are determined at an earlier stage; during embryogenesis.[6]

Subsequent studies widened the Hox related network that regulates the size but also identity of neuronal populations in the different segments of the Drosophila body. Until recently it was unknown how the competence to respond to a specific burst of AbdA expression with programmed cell death was restricted to late neuroblasts. A recent study showed that a late pNB transcription factor is responsible. Expression and function of AbdA is regulated by the transcription factor Grainyhead (Grh) in the abdomen. Grh is first expressed in later stages of neuroblast proliferation and is maintained through many subsequent divisions. It is responsible for specifying the regionalized neurogenesis patterns that are characteristic of postembryonic stages. In the thorax, Grh prolongs neural proliferation by maintaining mitotically active neuroblasts. On the contrary, Grh regulates region specific apoptosis in the abdomen by regulating the duration of the late phase pulse of expression of Grh as well as the competence of the neuroblasts to respond to it.[7] This study showed how a factor specific to late-stage neural progenitors can regulate the time at which neural proliferation stops, and identifies mechanisms linking it to the Hox axial patterning system.

The thoracic neuroblast lineage generates both neurons and glial cells via an asymmetric first cell division which segregates the glial factors prospero and glial cells missing (gcm) to the glial precursors. CycE, that is also asymmetrically expressed after the first division of thoracic neuroblast, functions upstream of *prospero* and *gcm* to specify the neuronal sublineage. These events do not require any Hox input and therefore the thoracic neuroblast lineage differentiation represents the default ground state. On the contrary, abdominal neuroblasts generate only glia cells via a symmetric first division which segregates the glial factors prospero and gcm to both daughter cells. This is made possible through a downregulation of a G1 cyclin, CycE, that occurs via the action of AbdA and AbdB. Loss of CycE function causes homeotic transformation of thoracic neuroblasts to abdominal neuroblasts whereas ectopic CycE induces the reverse transformation.[8] Other components of the cell cycle seem to have a minor role in this process, suggesting a critical role for CycE in regulating cell fate in segment-specific neural lineages.

Much remains to be learned by other factors that regulate neuroblast activity. In particular, some of the humoral factors and short-range 'niche' signals that modulate neuroblast activity during postembryonic development have been identified. These may well be also under the control of the Hox axial patterning system but this remain to be seen.

Hox Gene Function in Stem Cells during Mammalian Development

The developmental functions of Hox genes have been elucidated mainly through gene inactivation studies but also through transgenic animals where Hox gene expression domains were altered. The large number of these studies indicated that Hox gene products act in a region specific, combinatorial and partly redundant fashion to specify antero-posterior (AP) identities in all three germ layers of the developing embryo.[9,10] Many of the Hox mutations resulted in classic homeotic transformations where a part of the body develops into the likeness of another.[11,12] In these cases the precursors of the affected tissue were left expressing the wrong complement of Hox genes suggesting that Hox genes impart identity in a combinatorial manner. It was also apparent from these studies that the inherent genetic redundancy built in the Hox axial patterning system often masked the full spectrum of Hox activities. Intriguingly, in the few cases where Hox information was completely eliminated or stripped down to the bare minimum, cells were locked to an immature state and whole structures or lineages went missing.

Combined loss of function of the Drosophila labial homologues *Hoxa1* and *Hoxb1* in the developing hindbrain, rhombomere 4 (r4) in particular, is a case demonstrating this point. Since *Hoxb2* expression in r4 is dependent upon Hoxb1 function in the same territory[13] this genetic combination resulted essentially in a triple knock out leaving the presumptive r4 expressing only *Hoxa2*. The territory itself was not completely lost but there was a cell autonomous defect in r4 neural crest cell generation.[14] This in turn resulted in early involution of the second pharyngeal arch and loss of its derivative skeletal elements.[15] In contrast, other cell lineages such as motor neurons are generated in this territory albeit with changed identities and atypical behavior.[15] Interestingly, the labial Drosophila mutants displayed a similar phenotype in the tritocerebrum, a territory where only *labial* is expressed. The neural tritocerebral progenitor cells are present and correctly located in the mutant domain. Furthermore, the postmitotic progeny of these progenitors, the cells that are fated to become neurons in the wild type, are also generated and correctly located in the mutant tritocerebral domain and remain so throughout subsequent embryogenesis. However, these cells seem to remain in an undifferentiated state and do not express any of the numerous neuronal molecular markers that positionally equivalent neuronal cells express in the wild type neither do they extend axons or dendrites. In contrast to the absence of neuronal cell fate in the *labial* mutant domain, the generation of glial cells within this mutant domain appears to be unaffected.[16]

Targeted disruption of *Hoxa3* resulted in loss of the thymus—a tissue derived through the mesenchymal neural crest of the third pharyngeal arch and its interaction with surrounding arch and pouch. In these mutants neural crest is generated and migrates correctly but fails to differentiate and properly interact with the surrounding tissue to generate the thymus.[17] Combined loss of function of paralogous group 3 Hox genes lead to similar defects, namely the loss of an entire structure. Mice homozygous for either *Hoxa3* or *Hoxd3* mutations have no defects in common whereas combinations of the two mutations exacerbate the respective phenotypes.[17] *Hoxb3* mutants display only minor defects in the cervical vertebrae and the IX nerve. Strikingly, when either *Hoxa3* or *Hoxd3* loss of function mutations are combined with *Hoxb3* loss of function mutation complete loss of the entire atlas was observed.[18] This phenotype could be attributed to mesodermal progenitors or stem cells failing to expand and differentiate.

Loss-of-structure phenotypes were also observed in combined loss of function of *Hox* genes of the paralogous group 11. Combination of *Hoxa11* and *Hoxd11* loss of function mutations resulted in a dramatic loss of the radius and ulna of the forelimb that were nearly completely eliminated. This suggested that paralogous Hox genes act together not to simply pattern limbs along the proximo distal axis but, crucially, to direct growth as well.[19] Furthermore, in the triple *Hoxa11/Hoxd11/Hoxc11* loss of function mutants a complete loss of metanephric kidney induction was observed. In these mutants the metaneprhic blastem condensed but failed to maintain the *pax-eya-six* regulatory cascade and initiate subsequent metanephric induction.[20]

The cases described above demonstrate that patterning, lineage specification and cellular growth are linked, with Hox genes playing a central role in these processes. Proper AP specification, in the form of a specific Hox code, is necessary for stem and progenitor cells to expand, respond appropriately to signals and interact with surrounding tissues. The definition of Hox conferred AP specification in transcripteome and proteome terms will be crucial if we want to understand these processes and put this knowledge at work with the ultimate goal been in vitro tissue generation and in vivo regeneration.

Hox Genes and Hematopoietic Stem Cells

The initial studies of human *HOX* gene expression and function in hematopoiesis used immortalized cell lines that were mostly established from leukemias. The involvement of *Hox* genes in human leukemias is now indirectly supported by their observed aberrant expression,[21,22] chromosomal translocations,[23,24] chromosomal translocations involving their cofactor PBX1[25,26] and chromosomal rearrangements involving their upstream regulators MLL1[27,28] and CDX2.[29] Gene expression profiling by microarrays showed that Hox gene expression is deregulated in a multiplicity of leukemias.[30] In these studies, *HOXA9* expression was the one, among that of 6800 genes, in patients with acute myeloid leukemia[21] that correlated mostly with treatment failure.[30] However, direct evidence for the functional contribution of *Hox* genes in the development of leukemias mediated by E2A-PBX1[30,31] and by some *MLL* fusion genes has only recently been obtained.[32] During normal hematopoiesis *Hox* genes of the A, B, and C clusters are transcribed, with their expression being confined to primitive subpopulations[27,33,34] whereas expression of the *HOXD* cluster genes is rare in hematopoietic cell lines.[35] These observations have prompted several studies that examined the effect of HOX genes in HSC expansion, differentiation and lineage commitment using a variety of approaches. The results point to an extremely complex role of Hox genes in hematopoietic development and differentiation.

HSC Specification, Expansion and Self Renewal

Hox genes have been implicated in mediating HSC specification, expansion and self-renewal. *Hoxb4* inducible, timed expression confers definitive long-term lymphoid-myeloid engraftment potential on mouse embryonic stem cells without apparent preference for either lineage. Similarly, retroviral transduction of *Hoxb4* in yolk sac cells also promoted their switch to the definitive HSC phenotype. In this case, however, long term expression of *Hoxb4* favoured the myeloid lineage.[36] The ectopic expression of *HOXB4* in mouse and human bone marrow cells increases the self-renewal of hematopoietic stem cells[37] in vivo, without affecting subsequent lineage-specific differentiation.[38,39] More recent studies have also shown that *HOXB4* induced rapid ex vivo expansion of the transduced HSCs.[40] Strikingly, the same effect could be induced using a soluble recombinant HOXB4 protein.[37,41] Importantly, the ex vivo-HOXB4-expanded HSCs retained their normal differentiation and long-term repopulation potential, and no hematologic abnormalities have been detected in large groups of mice that received transplants with HOXB4-transduced HSCs.[42] Expansion of HSC population may not be a property unique to Hoxb4 as the leukemogenic gene *HOXA9* has similar potential to expand HSCs in the preleukemic phase of the disease. That provided intriguing evidence that the *Hox*-induced enhancement of HSC expansion and self renewal may extend to other *Hox* genes.[43] However, Hoxb4 is the only Hox gene identified so far capable of expanding the HSC population without affecting subsequent differentiation. Interestingly, in the mouse, combined deficiency in both *Hoxb3* and *Hoxb4* results in defects in endogenous hematopoiesis. This consisted of a significant reduction in the HSC pool which in turn resulted from a reduced proliferative capacity and slower cell cycle kinetics of the HSCs. Combined *Hoxb3* and *Hoxb4* deficiency did not alter the balance of lineage commitment.[44] These results directly implicate *Hoxb3* and *Hoxb4* in regulating in vivo stem cell regeneration and affording maximal proliferative response.

The Mll gene encodes a trithorax-group chromatin regulator essential for the development of HSCs during development. Chromosomal translocations disrupting the Mll gene result in leukemia with aberrant expression of some native target genes that include Hox genes.[45] In the absence of *Mll*, early hematopoietic progenitors develop despite reduced expression of Hoxa, Hoxb and Hoxc genes but have markedly reduced ability to generate hematopoietic colonies, a process requiring cell division and differentiation. Reactivation of a subset of Hox genes or, remarkably, of even a single Hox gene can rescue this proliferative defect in sharp contrast to other Mll target genes such as *Pitxa2* which fail to do so.[45] Similarly, zebrafish *cdx4* mutants fail to specify blood progenitors and this defect can be rescued by overexpression of *Hoxb7a* and *Hoxa9a* but not by *scl*.[46]

The identification of Hox genes as HSC regulators has been exploited to develop strategies to efficiently expand HSC ex vivo. This is a crucial step for the success of therapies based on HSC transplantation and the understanding of mechanisms underlying HSC self-renewal and proliferation. These advances should facilitate the development of strategies for the management of leukemia.

HSC Lineage Commitment

A number of functional studies have provided evidence that proteins encoded by *HOX* genes influence lineage commitment decisions of hematopoietic stem and progenitor cells.

Studies of *HOXB* expression in normal human bone marrow show that that *HOXB3*, *HOXB4* and *HOXB5* are easily detectable in HSCs.[22,27,47] Anti-sense inhibition of protein expression in normal human bone marrow has provided information about the function of a number of *HOXB* genes.[47] HSCs treated with anti-*HOXB3* oligonucleotides lose the ability to generate myeloid or erythroid colonies. In addition, anti-*HOXB4* and anti-*HOXB5* oligonucleotides produce dose-dependent decreases in myeloid and erythroid colony formation from HSCs. Culture conditions that promoted self-renewal induced a 3-10-fold increase in *HOXB4* expression among CD34+ bone marrow cells, while conditions promoting differentiation induced only a slight increase.[48] On the other hand, expression of *HOXB3* was dramatically reduced in CD34+CD38+ progenitors and almost undetectable in myeloid and erythroid progenitor populations[27] and CD34- cells.[22] *HOXB4* and *HOXB5* expression continued in CD34+CD38+ cells and myeloid progenitors, but both were downregulated in erythroid progenitors.[27] These studies suggest that *HOXB3*, *HOXB4* and *HOXB5* act in very early progenitors, possibly before myeloid and erythroid lineage restriction. Anti-*HOXB6* oligonucleotides reduced the formation of myeloid colonies from HSCs by about 75%, but they had no effect on erythroid colonies.[47] Although expression of the *HOXB6* gene has not been detected in bone marrow progenitors it has been observed in cultures that selectively induce granulocytic differentiation from CD34+ bone marrow cells.[47] The introduction of anti-*HOXB7* oligonucleotides inhibited myeloid colony formation in total human bone marrow.[49] In contrast to *HOXB6*, the expression of *HOXB7* is initiated in conditions that selectively induce monocytic but not granulocytic differentiation from the HL-60 cell line. Furthermore, granulocytic differentiation from the HL-60 cell line was inhibited by overexpression of *HOXB7*.[49] Thus, *HOXB6* and *HOXB7* appear to function later in hematopoietic differentiation to regulate the selection between granulocytic vs. monocytic differentiation with HOXB6 favoring the former and HOXB7 the latter.

During the progression of differentiation, the genes of the *HOXB* cluster appear to follow a general pattern of sequential activation and subsequent down regulation that corresponds with their 3' to 5' position on the chromosome. Additionally the functional studies summarized above provide evidence that the role of *HOXB* genes correlates with their location on the chromosome. In general, products of more 3' genes affect early differentiation and lineage restriction events, while proteins encoded by the more 5' genes tend to be lineage specific and function in lineage commitment. Fewer data are available on the expression and function of human *HOXA* and *HOXC* clusters but a similar pattern of expression and function for these genes may emerge.

Overexpression of *HOXA5* in CD34+ cord blood progenitors induced a shift toward myeloid and away from erythroid differentiation. Importantly, the greatest effects of forced *HOXA5* expression were seen in CD34+CD38- cells and in conditions that favored erythropoiesis over myelopoiesis.[50] Consistent with these observations, RT-PCR analysis of human bone marrow showed that *HOXA5* expression was easily detectable in HSCs and myeloid progenitors, but barely detectable in an erythroid progenitor population.[27] Similarly, overexpression of *HOXA10* in human CD34+ cord blood cells sharply reduced the in vitro production of erythroid,[51] T, B and NK cells[52] whereas it increased the production of myeloid cells.[51,52] Forced expression of HoxA10 in CD34+ human cells transplanted in SCID mice resulted in an increase of myeloid cells and a consequent reduction of B lineage cells.[51] Expression of *HOXA10* was detected in bone marrow HSC and myeloid progenitor populations but not mature neutrophils, monocytes or lymphocytes.[51,52] However, in contrast to *HOXA5,* transcripts for *HOXA10* were also readily detectable in erythroid progenitors.[27] These data suggest that both HOXA5 and HOXA10 proteins promote myeloid differentiation.

Members of the *HOXC* cluster of genes have been linked to both early hematopoietic events and exclusive terminal lymphoid differentiation. Overexpression of *HOXC4* in CD34+ cells has been shown to dramatically increase the proliferation of lymphoid as well as myeloid and erythroid precursors.[53] *HOXC4* transcripts have been detected in CD34+CD38+ progenitors and in terminally differentiated T and B cells[54] but not in mature granulocytes or monocytes.[55] Expression of *HOXC6* has been detected in mature T lineage cells, but not in CD34+ progenitors, suggesting that it functions later in lymphoid differentiation.

Proteins encoded by *Drosophila Polycomb group*[56] genes function as transcriptional repressors of *Hox* genes in mouse and human homologs of *Pc-G* genes are postulated to regulate *HOX* genes.[56] At least nine genes homologous to *Drosophila Pc-G* genes have been identified in human bone marrow.[56] RT-PCR analysis showed very low to negligible levels of expression of these homologs in primitive progenitors whereas expression increased steadily as differentiation progressed to myeloid and erythroid precursors, the only notable exception been BMI which was expressed in primitive progenitors and downregulated with differentiation. Upregulation of the *Pc-G* genes was not simultaneous; expression of some was initiated in progenitors while others were upregulated at later stages of differentiation.[56] Thus, in general the expression pattern observed for *Pc-G* genes appears to correlate inversely with that of *HOX* genes, which show a pattern of preferential expression in primitive progenitors.[22,27] The reciprocal and graded pattern observed for *Pc-G* and *HOX* gene expression is consistent with the hypothesis that proteins encoded by *Pc-G* genes play a role in the progressive 3' to 5' repression of *HOX* genes during hematopoietic differentiation and lineage commitment.[27,56]

Hox Genes and Cancer Stem Cells

Earlier models of carcinogenesis stipulated that uncontrolled proliferation arises as a result of serial acquisition of genetic lesions that lift normal restrictions, upregulate proliferative responses and abrogate programmed cell death. In addition to these models the recent stem cell model for carcinogenesis suggests that a key event would be the regulatory disruption of genes involved in the regulation of stem cell self renewal. There is evidence to suggest that Hox genes may be implicated in cell cycle control, differentiation as well as programmed cell death. Understanding their functions in this context may provide important diagnostic tools and means to control carcinogenesis.

The oncogenic potential of HOX genes has clearly been implicated in leukaemias[57] (see also below). Major changes in HOX gene expression have been detected in primary solid tumours (kidney, colon, and small cell lung cancer) and cell lines derived from them when compared with the corresponding normal adult organ.[58] In many cases, HOX genes that are normally only active during embryonic development are reexpressed in the neoplastic cells.[59] Such alterations involve HOX genes crucial for organ development as well as HOX genes apparently unrelated to normal organogenesis of the tissue. Links between HOX gene expression and malignant transformation have been initially investigated based on the hypothesis that genes expressed during embryogenesis, but not in adult tissues, can be reexpressed in neoplastic tissue. As it turns out this hypothesis is only partially supported by the data implying that it is an oversimplification of the role Hox genes play in carcinogenesis.[59,60] Besides the involvement of altered Hox gene expression in primary solid tumours, misexpression of HOX genes is detected in metastatic lesions with respect to the primary tumour of origin and the normal tissue suggesting an implication of homeoproteins in cancer evolution.[61] Furthermore, in some cases HOX gene expression appears to be linked to stage of tumour progression and histological tumour type.[62] Thus it is tempting to speculate that initial deregulation of Hox gene expression at the tissue stem cell level propagates itself through Hox cross-regulatory interactions endowing cells with metastatic potential and broadening their differentiation potential.

Excessive signalling may also lead to HOX gene upregulation and this in turn may accelerate the carcinogenic process. Autocrine hGH production by human mammary carcinoma cells increased the expression and transcriptional activity of HOXA1. This in turn resulted in up-regulation of *Bcl-2* and increased total cell numbers. HOXA1 also abrogated the apoptotic response of the mammary carcinoma cells to doxorubicin. Furthermore, overexpression of *HOXA1* in these cells markedly enhanced anchorage independent cell proliferation and caused their oncogenic transformation, rendering them capable of aggressive tumour formation.[63] The underlying assumption, which however has not been proven, is that Hox connected oncogenic activities are produced by wild type rather than mutant proteins as a by product of their normal activities in the wrong cellular context. For example, overexpression of HOXB7 activates bFGF which in turn promotes cellular proliferation. It is plausible that bFGF is a normal HOXB7 target during development and its inappropriate activation in melanoma and ovarian epithelial cells accelerates cellular proliferation and thus tumorigenesis.[64,65]

Direct interactions of Hox genes and their regulators with the machinery controlling cell cycle progression have been identified. The cell cycle regulator geminin associates transiently with members of the Hox-repressing polycomb complex, with the chromatin of Hox regulatory DNA elements and with Hox proteins. The interaction between geminin and Hox proteins prevents Hox proteins from binding to DNA, inhibits Hox-dependent transcriptional activation of reporter and endogenous downstream target genes. Geminin also directly interacts with Cdt1 which is a crucial component of the replication licencing machinery and this interaction prevents replication. Therefore geminin is coordinating developmental and proliferative control through competitive interactions. Thus, it is conceivable that aberrant levels of Hox homeoproteins may lead to unwarranted replication through sequestration of geminin and release of Cdt1 inhibition.[66] Timely destruction of HOXC10 by the anaphase promoting complex appears to be necessary for the orderly progression from metaphase to anaphase. HOXC10 degradation coincides with cyclinA degradation and is carried out by the ubiquitin-dependent proteasome pathway. When two destruction motifs in the protein are mutated HOXC10 is stabilised and cells accumulate in metaphase. This is another potential mechanism via which Hox genes constitute a link between developmental regulation and cell cycle control.[67] Yet another tentative link of Hox function with cell cycle regulation is the finding that HOXA5 regulates p53; expression of which protects against malignant transformation. Levels of p53 mRNA are low in a large proportion of breast tumours and breast cancer cell lines and patient tumours display a coordinate loss of HOXA5 mRNA and protein expression. The HOXA5 promoter region was methylated in 16 out of 20 p53-negative breast tumour specimens. Importantly, HOXA5 can bind to the p53 promoter and activate its transcription. Additionally, expression of HOXA5 in epithelial cancer cells expressing wild-type p53, but not in isogenic variants lacking the p53 gene, led to apoptotic cell death.[68] To further investigate this Chen et al engineered a *p53*-mutant breast cancer cell line to inducibly express *HOXA5*. Induction of*HOXA5* expression caused caspase 2- and caspase 8-mediated apoptotic cell death. Thus, loss of expression of p53 in human breast cancer appears be primarily due to lack of expression of HOXA5.[69] It will be of interest to learn whether other *HOX* genes have pro- or anti- apoptotic activity.

Malignancy is also associated with the failure of cells to terminally differentiate and several homeobox-containing genes have been found to be involved with terminal differentiation. Overexpression of some *HOX* genes is associated with loss of differentiation. For example, in human prostate cancer, overexpression of *HOXC8* is associated with the loss of tumour differentiation, suggesting that it is involved in the acquisition of the invasive and metastatic character of this malignancy.[70]

The evidence so far suggests that the gain or loss of function of Hox genes promotes tumorigenesis as a consequence of their inappropriate effects on growth and differentiation. However the current state of our knowledge is insufficient to understand the relationship between individual Hox genes and the specific cancer phenotype that their misexpression contributes to. Furthermore, an indisputable link between Hox gene misexpression in tissue specific stem cells and consequent carcinogenesis has not been forged. Perhaps the most promising venues would be to (a) characterise the Hox expression profiles of tissue specific stem cells in physiological and tumorigenic state and (b) explore this possibility in animal models through the controlled misexpression of Hox genes in specific tissue stem cells.

Conclusions

Patterning, lineage specification and cellular growth are linked and it is becoming obvious that Hox genes are playing a central role in these processes. Proper AP specification, in the form of a specific Hox code, is necessary for stem and progenitor cells to expand, respond appropriately to signals and interact with surrounding tissues. It is now accepted that Hox functionality is context dependent and restricted by the developmental history of the cell as well as the signals the cell receives. Understanding the role of Hox genes in specific stem cell contexts holds the promise of harnessing their potential for in vivo tissue regeneration and in vitro tissue generation. To this end we will need to specify their context dependent activity in terms of transcripteome and proteome changes. Already, the identification of Hox genes as HSC regulators has been exploited to develop strategies to efficiently expand HSC ex vivo. This is a crucial step for the success of therapies based on HSC transplantation and the understanding of mechanisms underlying HSC self-renewal and proliferation. These advances should facilitate the development of strategies for the management of leukemia. Similar advances on the role of Hox genes in maintenance, expansion and differentiation of other cancer stem cells may give us better diagnostic and therapeutic tools.

References

1. Betschinger J, Knoblich JA. Dare to be different: Asymmetric cell division in Drosophila, C. elegans and vertebrates. Curr Biol 2004; 14(16):R674-685.
2. Peterson C, Carney GE, Taylor BJ et al. reaper is required for neuroblast apoptosis during Drosophila development. Development 2002; 129(6):1467-1476.
3. White K, Grether ME, Abrams JM et al. Genetic control of programmed cell death in Drosophila. Science 1994; 264(5159):677-683.
4. Bello BC, Hirth F, Gould AP. A pulse of the Drosophila Hox protein Abdominal-A schedules the end of neural proliferation via neuroblast apoptosis. Neuron 2003; 37(2):209-219.
5. Lohmann I, McGinnis N, Bodmer M et al. The Drosophila Hox gene deformed sculpts head morphology via direct regulation of the apoptosis activator reaper. Cell 2002; 110(4):457-466.
6. Prokop A, Bray S, Harrison E et al. Homeotic regulation of segment-specific differences in neuroblast numbers and proliferation in the Drosophila central nervous system. Mech Dev 1998; 74(1-2):99-110.
7. Cenci C, Gould AP. Drosophila Grainyhead specifies late programmes of neural proliferation by regulating the mitotic activity and Hox-dependent apoptosis of neuroblasts. Development 2005; 132(17):3835-3845.
8. Berger C, Pallavi SK, Prasad M et al. A critical role for cyclin E in cell fate determination in the central nervous system of Drosophila melanogaster. Nat Cell Biol 2005; 7(1):56-62.
9. Trainor PA, Krumlauf R. Hox genes, neural crest cells and branchial arch patterning. Curr Opin Cell Biol 2001; 13(6):698-705.
10. Favier B, Dolle P. Developmental functions of mammalian Hox genes. Mol Hum Reprod 1997; 3(2):115-131.
11. Horan GS, Ramirez-Solis R, Featherstone MS et al. Compound mutants for the paralogous hoxa-4, hoxb-4, and hoxd-4 genes show more complete homeotic transformations and a dose-dependent increase in the number of vertebrae transformed. Genes Dev 1995; 9(13):1667-1677.
12. Rijli FM, Mark M, Lakkaraju S et al. A homeotic transformation is generated in the rostral branchial region of the head by disruption of Hoxa-2, which acts as a selector gene. Cell 1993; 75(7):1333-1349.
13. Maconochie MK, Nonchev S, Studer M et al. Cross-regulation in the mouse HoxB complex: The expression of Hoxb2 in rhombomere 4 is regulated by Hoxb1. Genes Dev 1997; 11(14):1885-1895.
14. Gavalas A, Trainor P, Ariza-McNaughton L et al. Synergy between Hoxa1 and Hoxb1: The relationship between arch patterning and the generation of cranial neural crest. Development 2001; 128(15):3017-3027.

15. Gavalas A, Studer M, Lumsden A et al. Hoxa1 and Hoxb1 synergize in patterning the hindbrain, cranial nerves and second pharyngeal arch. Development 1998; 125(6):1123-1136.
16. Hirth F, Hartmann B, Reichert H. Homeotic gene action in embryonic brain development of Drosophila. Development 1998; 125(9):1579-1589.
17. Manley NR, Capecchi MR. The role of Hoxa-3 in mouse thymus and thyroid development. Development 1995; 121(7):1989-2003.
18. Manley NR, Capecchi MR. Hox group 3 paralogous genes act synergistically in the formation of somitic and neural crest-derived structures. Dev Biol 1997; 192(2):274-288.
19. Davis AP, Witte DP, Hsieh-Li HM et al. Absence of radius and ulna in mice lacking hoxa-11 and hoxd-11. Nature 1995; 375(6534):791-795.
20. Wellik DM, Hawkes PJ, Capecchi MR. Hox11 paralogous genes are essential for metanephric kidney induction. Genes Dev 2002; 16(11):1423-1432.
21. Afonja O, Smith Jr JE, Cheng DM et al. MEIS1 and HOXA7 genes in human acute myeloid leukemia. Leuk Res 2000; 24(10):849-855.
22. Kawagoe H, Humphries RK, Blair A et al. Expression of HOX genes, HOX cofactors, and MLL in phenotypically and functionally defined subpopulations of leukemic and normal human hematopoietic cells. Leukemia 1999; 13(5):687-698.
23. Barbouti A, Hoglund M, Johansson B et al. A novel gene, MSI2, encoding a putative RNA-binding protein is recurrently rearranged at disease progression of chronic myeloid leukemia and forms a fusion gene with HOXA9 as a result of the cryptic t(7; 17)(p15; q23). Cancer Res 2003; 63(6):1202-1206.
24. Lam DH, Aplan PD. NUP98 gene fusions in hematologic malignancies. Leukemia 2001; 15(11):1689-1695.
25. Hunger SP, Galili N, Carroll AJ et al. The t(1; 19)(q23; p13) results in consistent fusion of E2A and PBX1 coding sequences in acute lymphoblastic leukemias. Blood 1991; 77(4):687-693.
26. Kamps MP, Baltimore D. E2A-Pbx1, the t(1; 19) translocation protein of human pre-B-cell acute lymphocytic leukemia, causes acute myeloid leukemia in mice. Mol Cell Biol 1993; 13(1):351-357.
27. Sauvageau G, Lansdorp PM, Eaves CJ et al. Differential expression of homeobox genes in functionally distinct CD34+ subpopulations of human bone marrow cells. Proc Natl Acad Sci USA 1994; 91(25):12223-12227.
28. Ernst P, Wang J, Korsmeyer SJ. The role of MLL in hematopoiesis and leukemia. Curr Opin Hematol 2002; 9(4):282-287.
29. Rawat VP, Cusan M, Deshpande A et al. Ectopic expression of the homeobox gene Cdx2 is the transforming event in a mouse model of t(12; 13)(p13; q12) acute myeloid leukemia. Proc Natl Acad Sci USA 2004; 101(3):817-822.
30. Golub TR, Slonim DK, Tamayo P et al. Molecular classification of cancer: Class discovery and class prediction by gene expression monitoring. Science 1999; 286(5439):531-537.
31. Thorsteinsdottir U, Krosl J, Kroon E et al. The oncoprotein E2A-Pbx1a collaborates with Hoxa9 to acutely transform primary bone marrow cells. Mol Cell Biol 1999; 19(9):6355-6366.
32. Ayton PM, Cleary ML. Transformation of myeloid progenitors by MLL oncoproteins is dependent on Hoxa7 and Hoxa9. Genes Dev 2003; 17(18):2298-2307.
33. Giampaolo A, Pelosi E, Valtieri M et al. HOXB gene expression and function in differentiating purified hematopoietic progenitors. Stem Cells 1995; 13(Suppl 1):90-105.
34. Pineault N, Helgason CD, Lawrence HJ et al. Differential expression of Hox, Meis1, and Pbx1 genes in primitive cells throughout murine hematopoietic ontogeny. Exp Hematol 2002; 30(1):49-57.
35. Lawrence HJ, Sauvageau G, Humphries RK et al. The role of HOX homeobox genes in normal and leukemic hematopoiesis. Stem Cells 1996; 14(3):281-291.
36. Kyba M, Perlingeiro RC, Daley GQ. HoxB4 confers definitive lymphoid-myeloid engraftment potential on embryonic stem cell and yolk sac hematopoietic progenitors. Cell 2002; 109(1):29-37.
37. Amsellem S, Pflumio F, Bardinet D et al. Ex vivo expansion of human hematopoietic stem cells by direct delivery of the HOXB4 homeoprotein. Nat Med 2003; 9(11):1423-1427.
38. Sauvageau G, Thorsteinsdottir U, Eaves CJ et al. Overexpression of HOXB4 in hematopoietic cells causes the selective expansion of more primitive populations in vitro and in vivo. Genes Dev 1995; 9(14):1753-1765.
39. Buske C, Feuring-Buske M, Abramovich C et al. Deregulated expression of HOXB4 enhances the primitive growth activity of human hematopoietic cells. Blood 2002; 100(3):862-868.
40. Antonchuk J, Sauvageau G, Humphries RK. HOXB4-induced expansion of adult hematopoietic stem cells ex vivo. Cell 2002; 109(1):39-45.
41. Krosl J, Austin P, Beslu N et al. In vitro expansion of hematopoietic stem cells by recombinant TAT-HOXB4 protein. Nat Med 2003; 9(11):1428-1432.
42. Krosl J, Beslu N, Mayotte N et al. The competitive nature of HOXB4-transduced HSC is limited by PBX1: The generation of ultra-competitive stem cells retaining full differentiation potential. Immunity 2003; 18(4):561-571.

43. Thorsteinsdottir U, Mamo A, Kroon E et al. Overexpression of the myeloid leukemia-associated Hoxa9 gene in bone marrow cells induces stem cell expansion. Blood 2002; 99(1):121-129.
44. Bjornsson JM, Larsson N, Brun AC et al. Reduced proliferative capacity of hematopoietic stem cells deficient in Hoxb3 and Hoxb4. Mol Cell Biol 2003; 23(11):3872-3883.
45. Ernst P, Mabon M, Davidson AJ et al. An Mll-dependent Hox program drives hematopoietic progenitor expansion. Curr Biol 2004; 14(22):2063-2069.
46. Davidson AJ, Ernst P, Wang Y et al. cdx4 mutants fail to specify blood progenitors and can be rescued by multiple hox genes. Nature 2003; 425(6955):300-306.
47. Giampaolo A, Sterpetti P, Bulgarini D et al. Key functional role and lineage-specific expression of selected HOXB genes in purified hematopoietic progenitor differentiation. Blood 1994; 84(11):3637-3647.
48. Giannola DM, Shlomchik WD, Jegathesan M et al. Hematopoietic expression of HOXB4 is regulated in normal and leukemic stem cells through transcriptional activation of the HOXB4 promoter by upstream stimulating factor (USF)-1 and USF-2. J Exp Med 2000; 192(10):1479-1490.
49. Lill MC, Fuller JF, Herzig R et al. The role of the homeobox gene, HOX B7, in human myelomonocytic differentiation. Blood 1995; 85(3):692-697.
50. Crooks GM, Fuller J, Petersen D et al. Constitutive HOXA5 expression inhibits erythropoiesis and increases myelopoiesis from human hematopoietic progenitors. Blood 1999; 94(2):519-528.
51. Buske C, Feuring-Buske M, Antonchuk J et al. Overexpression of HOXA10 perturbs human lymphomyelopoiesis in vitro and in vivo. Blood 2001; 97(8):2286-2292.
52. Taghon T, Stolz F, De Smedt M et al. HOX-A10 regulates hematopoietic lineage commitment: Evidence for a monocyte-specific transcription factor. Blood 2002; 99(4):1197-1204.
53. Daga A, Podesta M, Capra MC et al. The retroviral transduction of HOXC4 into human CD34(+) cells induces an in vitro expansion of clonogenic and early progenitors. Exp Hematol 2000; 28(5):569-574.
54. Bijl J, van Oostveen JW, Kreike M et al. Expression of HOXC4, HOXC5, and HOXC6 in human lymphoid cell lines, leukemias, and benign and malignant lymphoid tissue. Blood 1996; 87(5):1737-1745.
55. Lawrence HJ, Stage KM, Mathews CH et al. Expression of HOX C homeobox genes in lymphoid cells. Cell Growth Differ 1993; 4(8):665-669.
56. Lessard J, Baban S, Sauvageau G. Stage-specific expression of polycomb group genes in human bone marrow cells. Blood 1998; 91(4):1216-1224.
57. Pineault N, Abramovich C, Ohta H et al. Differential and common leukemogenic potentials of multiple NUP98-Hox fusion proteins alone or with Meis1. Mol Cell Biol 2004; 24(5):1907-1917.
58. Cillo C. HOX genes in human cancers. Invasion Metastasis 1994; 14(1-6):38-49.
59. Abate-Shen C. Deregulated homeobox gene expression in cancer: Cause or consequence? Nat Rev Cancer 2002; 2(10):777-785.
60. Cillo C, Cantile M, Faiella A et al. Homeobox genes in normal and malignant cells. J Cell Physiol 2001; 188(2):161-169.
61. De Vita G, Barba P, Odartchenko N et al. Expression of homeobox-containing genes in primary and metastatic colorectal cancer. Eur J Cancer 1993; 29A(6):887-893.
62. Tiberio C, Barba P, Magli MC et al. HOX gene expression in human small-cell lung cancers xenografted into nude mice. Int J Cancer 1994; 58(4):608-615.
63. Zhang X, Zhu T, Chen Y et al. Human growth hormone-regulated HOXA1 is a human mammary epithelial oncogene. J Biol Chem 2003; 278(9):7580-7590.
64. Care A, Silvani A, Meccia E et al. HOXB7 constitutively activates basic fibroblast growth factor in melanomas. Mol Cell Biol 1996; 16(9):4842-4851.
65. Naora H, Montz FJ, Chai CY et al. Aberrant expression of homeobox gene HOXA7 is associated with mullerian-like differentiation of epithelial ovarian tumors and the generation of a specific autologous antibody response. Proc Natl Acad Sci USA 2001; 98(26):15209-15214.
66. Luo L, Yang X, Takihara Y et al. The cell-cycle regulator geminin inhibits Hox function through direct and polycomb-mediated interactions. Nature 2004; 427(6976):749-753.
67. Gabellini D, Colaluca IN, Vodermaier HC et al. Early mitotic degradation of the homeoprotein HOXC10 is potentially linked to cell cycle progression. EMBO J 2003; 22(14):3715-3724.
68. Raman V, Martensen SA, Reisman D et al. Compromised HOXA5 function can limit p53 expression in human breast tumours. Nature 2000; 405(6789):974-978.
69. Chen H, Chung S, Sukumar S. HOXA5-induced apoptosis in breast cancer cells is mediated by caspases 2 and 8. Mol Cell Biol 2004; 24(2):924-935.
70. Waltregny D, Alami Y, Clausse N et al. Overexpression of the homeobox gene HOXC8 in human prostate cancer correlates with loss of tumor differentiation. Prostate 2002; 50(3):162-169.

Chapter 9

Deregulation of the Hox Gene Network and Cancer

Clemente Cillo*

Abstract

Although the Hox genes have been identified as master regulatory genes controlling embryonic development, an alternative view on the role of the Hox gene network suggests that it regulates crucial processes at cellular level in eukaryotic organisms. Our working hypothesis considers the Hox network, at the nuclear cell level, as a decoding system for external inductive signals to activate specific genetic programs. We thus identify a cancer as an anomalous structure growing inside the human body and following, from a cellular and architectural viewpoint, the rules controlling body shape during embryonic development. In this chapter we will describe, according to present data, how the Hox gene network acts in specific types of human solid tumours (breast, prostate, bladder, kidney). Furthermore we identify three research areas potentially able to produce, in the near future, important achievements to increase our understanding of (i) the function of the Hox gene network; (ii) the identification of the HOX target genes; (iii) the molecular basis of human cancers.

Introduction

Although the Hox genes were originally identified as master regulatory genes controlling segment identity along the antero-posterior axis of animals during embryonal development—thoracic Hox gene mutations cause homeotic transformations of the vertebrae (see elsewhere in this book)—an alternative view of the role of Hox gene network as a whole suggests that it regulates crucial processes at cellular level in eukaryotic organisms. The function of the Hox network must be so crucial as to require the original Hox locus to become tetraploid during evolution while retaining its peculiar gene organisation across so many animal species.[1] We have proposed that the Hox network acts, at the nuclear level of eukaryotic cells, as a decoding system for external inductive signals that allow the activation of specific genetic programs. Specific genes in the network act as signal collectors (from growth factors and signal transduction pathway). These signals are transferred to key genes in the decoding network (decoder genes may vary in different cell phenotypes). Through transmitter genes, a response is sent outside the Hox network. This activates specific programs of effector genes (morphogenetic molecules, cell-cycle related proteins, apoptotic pathway) in order to achieve the changes induced by the signal received.[2] Multiple cellular processes are regulated through the Hox network: the acquisition and maintenance of spatial and temporal

*Correspondence to: Clemente Cillo—Department of Clinical and Experimental Medicine, Federico II University Medical School Via S. Pansini 5, 80131 Naples, Italy. Email: clecillo@unina.it. Present Address: Institute of Pathology, Molecular Pathology Division, University of Basel, Schonbeinstrasse 40, 4003 Basel, Switzerland. Email: ccillo@uhbs.ch

HOX Gene Expression, edited by Spyros Papageorgiou. ©2007 Landes Bioscience and Springer Science+Business Media.

cell allocation; the establishment of cell identity, achieving specific cell phenotypes through the decoding of microenvironment signals; the control of cell growth and proliferation through the interaction with cell-cycle and the apoptotic pathway; the process of cell-cell communication through cross-talk with morphogenetic molecules, growth factors and cytokines and signal transduction pathways; the antero-posterior patterning during embryonic development; the regulation of cell compartmentalisation and architectural organisation during morphogenesis and organognesis; and the rules that coordinate cell activity with evolutionary constraints such as number of digits in vertebrate and number of neck vertebrae in mammals.[3] We and other have thus postulated the determination of phenotype cell identity as the primary function of the Hox gene network distinct from the consequent function of controlling body segment identity.[4-6]

In order to test the robustness of the HOX gene network in controlling phenotype cell identity we have designed the following experiment on human prostate cells. We detect the expression patterns of the HOX gene network in a non tumorigenic epithelial cell line EPN[7] as well as in a cell line, EPN-PKM3, an EPN clone bearing a Pyk2 kinase-negative mutant (kindly provided by D. Tramontano—Federico II University Medical School, Naples). Pyk2 is a cytoplasmic tyrosine kinase, related to focal adhesion kinases, actively expressed in human prostate epithelia and inversely correlated, through its expression, to prostate malignancy. Although the patterns of expression relative to the HOX gene network are undistinguishable in the two cell phenothypes, cAMP treatment generates phenotype specific alterations on HOX gene expression as well as phenotype specific morphological and functional differences in EPN versus EPN-PKM3 cells (see prostate cancer section). A short cAMP exposure (24h) induces modification in the expression of five HOX genes in EPN cells (Fig. 1 left) whereas a single gene (HOX D9) becomes active in EPN-PKM3 cells (Fig. 1 right). Subtraction of Pyk 2 kinase in EPN-PKM3 seems to generate a delay in the modification of HOX gene expression induced by cAMP treatment. This suggests an interaction between Pyk 2, cAMP and the HOX gene network as well as the involvement of the HOX network in cAMP induced phenotype modifications in both epithelial cell phenotypes. Increasing the exposure to cAMP (72h) and in parallel to the increased morphological and functional neuroendocrine modifications of these cell phenotypes, the alterations in the expression of the HOX gene network concomitantly increases: 10 HOX genes alter their expression in EPN and 8 genes in EPN-PKM3 (Fig. 1). These modifications in the HOX gene expression are only partially shared (6 HOX genes) by the two cell phenotypes. Thus, prolonged exposure to cAMP induces differences in the HOX genes which modify their expression in the two epithelial prostate cell phenotypes. This suggests a direct interaction between Pyk 2 and specific HOX genes. One of the two altered HOX genes after cAMP in EPN-PKM3 cells is HOX A7, recently described as transcriptionally interacting with Pyk2 during HL-60 monocytic differentiation.[8]

Thus, the expression of the HOX gene network after exposure to cAMP, not only discriminates between two epithelial prostate cell phenotypes displaying a single gene difference but also allows identification of specific HOX genes able to interact with the gene, the focal adhesion kinase Pyk-2, for which the two cell phenotypes differ.

In the light of this way of thinking, fifteen years ago we started a project intended to study the role of the HOX gene network in tumour progression. This allowed us to suggest the involvement of HOX genes in several types of human cancers. In recent years, the increasing reliability of technologies for whole-genome RNA expression and the coupling of bioinformatics and gene expression has made it possible to compare, on large scale screenings, gene expression to be used for clinical evaluations. This has produced the definitive acquisition of HOX genes as "emerging stars in cancer research".[9] In this chapter we will describe, according to present results, how the HOX gene network acts in specific types of human solid cancers (breast, prostate, bladder, kidney) starting from our (now outdated) working procedure and comparing it with recent high-throughput molecular pathology

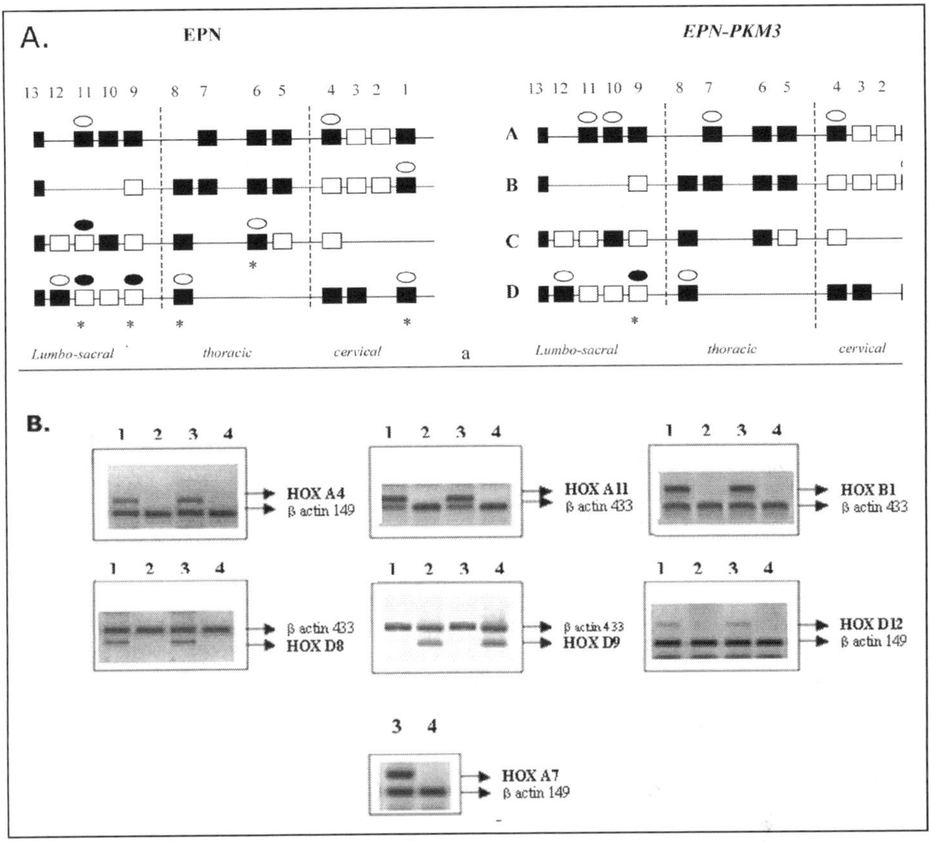

Figure 1. A) Schematic representation of the HOX gene network (see text for details). Expression of the whole HOX gene network detected by RT-PCR in EPN and EPN-PKM3 prostate cell lines: black and white squares indicate active or silent HOX genes, respectively. Asterisks indicate HOX genes altered after 24h cAMP exposure in EPN and EPN-PKM3 cells.?Black and white ovals indicate active or silent HOX genes altered in EPN and EPN-PKM3 cells after 72h cAMP exposure. Altered HOX genes are active or silent, in the same sense, after 24 and 72h cAMP exposure. B) RT-PCR expression of HOX A4, HOX A7, HOX A11, HOX B1, HOX D8 HOX D9 and HOX D12 in EPN (lane 1), EPN + 72h cAMP (lane 2), EPN-PKM3 (lane 3) and EPN-PKM3 + 72h cAMP (lane 4). Control coamplification of HOX A4, HOX D12 and HOX A7 with a 149 bp β-actin primer is reported. Control coamplification of HOX A11, HOX B1, HOX D8, HOX D9 and with a 433 bp β-actin primer is reported. Duplex PCR products were separated by ethidium 1.2% agarose gel electrophoresis.

analyses to show how specific HOX genes can be used as diagnostic and predictive tools in cancer research.[10]

Breast Cancer

Initial observations on *HOX* gene involvment in breast cancer already pointed out the role of genes from loci A and C of the HOX network.[11] Hox A1 gene has been implicated in mouse and human breast epithelial cell differentiation mostly through its regulation by retinoic acid[12] and steroids,[13] and its link to the expression of laminin.[14] HOX C6 has been detected in breast carcinomas as being able to contribute to breast cell phenotype through its interaction with other HOX genes (HOX B7) or repress their target genes.[15] Increased HOX D3 expression has been connected to angiogenesis through the regulation of αv β3 integrin and urokinase plasminogen activator (u-PA).[16] Transduction of breast carcinoma cells with HOX B7 induces bFGF expression and increases in vitro breast cells malignancy.[17] Recently, it has been shown that, in breast carcinomas, HOX A5 behaves as a transcriptional regulator of multiple target genes, two of which are p53[18] and the progesterone receptor.[19] Paralogous group 9 Hox genes (Hox a9, Hox b9 and Hox d9) act together in mutant mice controlling the differentiation of their mammary epithelial ductal system during pregnancy.[20]

We have recently described the expression of the complete HOX gene network in normal human breast and in primary breast carcinomas in order to understand whether the HOX network is implicated in breast cancer evolution and to identify, inside the network, the specific HOX genes primarily involved with breast tumorigenesis.[21] Our results show that the HOX gene network displays an overall expression pattern characteristic of normal breast tissue. Furthermore, the expression of thoracic HOX genes is similar in normal and neoplastic breast tissue indicating that these genes may be involved in breast organogenesis. In contrast, cervical and lumbo-sacral HOX genes manifest altered expression in primary breast cancers with respect to normal breast, which supports their involvement in breast cancer progression.

According to our results, specific HOX genes, such as HOX B2, HOX D3 and HOX D4 in the cervical part of the network and HOX D10, HOX A11 and HOX B13 in the lumbo-sacral part of the network, are good candidate genes to be tested in a large study to confirm their role in breast cancer progression. These observations stress the importance of the HOX network in cancer evolution and the possibility to target the HOX genes for future cancer therapies.

Tamoxifen significantly reduces tumour recurrence in early-stage estrogen-positive breast cancers. Approximately 40% of ER + breast cancers fail to respond or eventually develop resistance to tamoxifen, leading to disease progression. Current clinicopathological features including tumour stage and grade, and ERBB2 and EGFR expression fail to accurately identify individuals who are at risk for tumour recurrence. A Californian team[22] has recently reported the gene expression profiling (22,000 genes) of tumours from 60 women uniformly treated with adjuvant tamoxifen monotherapy. This cohort included 46% tamoxifen recurrences and 54% tamoxifen non recurrences with a 5-year follow-up-available after biopsy and matching of recurrences and nonrecurrent cases with respect to TNM staging[23] and tumor grade.[24] By an initial screening the authors demonstrated the existence of statistically significant differences in gene expression between the primary breast cancers of tamoxifen nonrecurrences and recurrences through the identification of 19 differentially expressed genes. By means of a more refined analysis using laser-capture microdissection (LCM) of tumour cells within each tissue section[25] they identified 9 differentially expressed genes. Three genes were identified as differentially expressed in both whole tissue sections and LCMs: HOX B13, the interleukin 17B receptor and EST A1240933. HOX B13 was hyperexpressed in tamoxifen recurrence cases, whereas IL17BR and A1240933 were hyperexpressed in tamoxifen nonrecurrence cases. These three genes, thus, predict the clinical outcome of patients treated with adjuvant tamoxifen. Due to the opposite patterns of expression between HOX B13 and IL17BR the authors then tested the possibility that the expression ratio of HOX B13 over IL17BR can be predictive of tamoxifen response, and compared the HOX B13:IL17BR ratio to well-established prognostic factors for breast cancer, such as patient age, tumor size, grade, and lymph node status PGR and ERBB2 expression. Results demonstrated that the expression ratio of HOX B13:IL17BR is a strong, independent predictor of treatment outcome in the setting of adjuvant tamoxifen therapy. Thus HOX B13, one of the genes located in the lumbo-sacral part of the HOX network which we have identified as a good candidate gene to be tested in a large study to confirm its role in breast cancer progression, has been validated through this study.

Bladder Cancer

Bladder carcinogenesis remains unclear despite the identification of chemical, environmental and genetic factors. Various genetic abnormalities have been identified in bladder cancer (usually transitional cell carcinoma or TCC):[26] an initial event on chromosome 9 (9p21 and 9p34), although the specific genes involved are still unidentified, causes the highest mutation index detectable in TCC.[27] Oncosuppressor genes linked to apoptosis and the cell cycle, such as *p53*[28] and *Rb,*[29] and oncogenes involved in signal transduction, such as *Ras*[30] have been connected to TCC. Finally, a transcription factor linked to embryonal development (*PAX 5*) and displaying oncogenic properties is localised in a key position (chromosome 9p13) in bladder carcinogenesis and seems to play an important role in TCC when inappropriately expressed.[31]

We have recently described the expression of the whole HOX gene network in pairs of normal-tumour bladder and in isolated tumor biopsies. Comparison between normal urothelium and bladder tumour has identified dramatic variations of expression in a block of three genes (HOX C4, HOX C5 and HOX C6) localized in the HOX C locus on the chromosome 12q13 and in the paralogous group 11 HOX genes, involved during normal development in the formation of the urogenital system. In particular, the *HOX C6* gene, is always silent in normal bladder but active in 100% of the 30 cases of TCC examined.[32] The physical location of the *HOX C* locus on chromosome 12q13 is within a chromosomal region recently described as amplified in bladder cancer.[33] This region contains (in addition to the entire HOX C locus) around sixty genes, including those which play a better documented oncogenic role such as *MDM2* (murine double minutes), *CDK4* (cyclin-dependent kinase 4) and *GLI* (glioblastoma-associated oncogene). Also among these genes are the genes coding for the cytokeratins *CK8* along with the cytokeratin genes *CK1, CK3, CK4, CK5, CK6A* and *CK6B.*[34] In addition, in physical contiguity at the 3' end of the *HOX C* locus, are located the genes for *CK7* and the basic hair Keratin 1 and 6, transcriptionally regulated by the *HOX C13* gene.[35] Another group of cytokeratin genes is located on chromosome 17p21.3[36] in physical contiguity with the *HOX B* locus. In bladder cancer, alteration of the expression of a series of CKs normally active in the urothelium, such as CK8, CK18, and CK 19, has been described, making it possible to suggest a distinction between TCCs through cytokeratin expression patterns.[37] In biology the concept of common functions performed by physically and/or evolutionarily contiguous genes is gaining more and more support.[38]

During normal development, paralogous group 11 *HOX* genes are involved in the formation of the posterior region, including the uro-genital system[39] (see kidney cancer section). According to our results, the *HOX A11* gene, active in most normal bladder tissue, tends to become silent in cancer tissue. The *HOX C11* gene, which is silent in normal urothelium, becomes active in almost all the TCC samples tested. Finally, the *HOX D11* gene, always silent in normal bladder tissue, appears heterogeneously expressed in bladder cancer. We, thus, hypothesize a role of HOX C6 and paralogous group 11 *HOX* genes in the regulation of the cell phenotypes involved with urological carcinogenesis.

Prostate Cancer

From the biological viewpoint, the heterogeneity of prostate cancers suggests the existence of multiple, independently-evolving, neoplastic foci, with severe limitations for depicting the molecular mechanisms of cancer progression. The introduction of laser-capture microscopy has facilitated the analysis of individual neoplastic loci. Gleason's formulation of a score system, used by pathologists as a prognostic indicator, and based on the sum of the two most prevalent grades of neoplastic foci makes it possible to identify higher Gleason grades as more advanced carcinomas.[40]

Early detection through prostate specific antigen (PSA) confines most tumours to prostate gland at presentation and provides an opportunity for curative surgery. However, up to 30% of patient having radical prostatectomy will relapse as a result of micrometastases dissemination at the time of surgery.[41] Thus, it will be crucial to identify molecular markers that define tumours at risk of relapse. Out of several candidate genes (loss of p53, myc amplification, loss of p27, loss of PTEN) neither combination nor single genes have been shown to be of such prognostic utility as to warrant clinical implementation.

Recently the gene expression patterns from 52 human prostate tumours and 50 normal prostate specimens, using oligonucleotide microarrays containing probes for approximately 12600 genes were studied in order to understand whether such patterns could be predictive of common clinical and pathological phenotypes relevant to patients' treatment.[42] The analysis indicated that 317 genes were hyper-expressed in prostate tumour samples and 139 genes in normal prostate samples. The expression patterns of the 52 tumours were further analysed in order to predict differences in the clinical behaviour of prostate cancers. No statistically significant gene expression correlate of the clinical and pathological features were observed with the exception of the Gleason score. A gene expression signature of GS was thus detectable: fifteen hyper-expressed genes and fourteen hypo-expressed genes correlate with GS and their expression is reproducible. While no single gene was statistically associated to recurrence, a 5 gene model predicts recurrence with 90% accuracy. The 5 genes are chromogranin A, platelet-derived growth factor receptor β (PDGFRβ), HOX C6, inositol triphosphate receptor 3 (IPTR3) and sialyltransferase-1. These data support the notion that the clinical behaviour of prostate cancer is linked to underlying differences in gene expression, detectable at the time of diagnosis. Thus, a 5 gene expression model alone accurately predicts patient outcome following prostatectomy and one of these genes is HOX C6, the gene we have described as inactive in 100% of bladder cancers. Furthermore, the association of specific genes' expression, such as PDGFRβ or HOX C6, with outcome raises the possibility that expression analysis may prove useful in selecting patients for emerging mechanism-based therapeutics.

Another study has shown the upregulation of genes from the HOX C locus (HOX C4, HOX C5, HOX C6 and HOX C8) in malignant prostate cell lines and lymph node metastases.[43] The HOX network expression patterns of lymph node metastases and cell lines derived from lymph node metastases are very similar, whereas patterns of HOX gene expression are distinct when detected in benign cells or malignant cell lines derived from metastases of prostate adenocarcinoma in other organ sites. Laser capture microdissection and examination of pairs tumour-normal prostate biopsies confirmed overexpression of the HOX C genes in primary prostate tumours. Overexpression of HOX C8 in LNCaP cells suppresses transactivation by androgen receptors suggesting the possibility that HOX C hyper-expression predisposes tumour cells to androgen independency. Thus, HOX C locus genes and HOX C6 in particular appear to play a major role in prostate cancer.

In the light of these observations we have recently investigated the involvement of the HOX gene network in neurendocrine differentiation of human advanced prostate cancer.[44] We have set up a model of human prostate tumour evolution by using: (a) primary cultures of prostate fibroblasts, as for the mesenchymal compartment; (b) a nontumorigenic prostate epithelial cell line (EPN); (c) three malignant cell lines generated from metastasis of prostate adenocarcinomas with different bodily localization: bone (PC3), brain (DU145), and lymph-node (LNCaP), and characterized by androgen independency (PC3 and DU145) or dependency (LNCaP). The expression patterns of the HOX gene network in these human prostate cell phenotypes, representing different stages of prostate physiology and prostate cancer progression, make it possible to discriminate between the different human prostate cell lines and to identify loci and paralogous groups harbouring the HOX genes mostly involved in prostate organogenesis and cancerogenesis.

cAMP is known to induce epithelial-neurocrine differentiation in prostate cancer cells.[45] On the basis of this knowledge we have investigated the effect of sustained intracellular level cAMP on the HOX gene network expression. cAMP induces a substantial difference in the expression of the HOX gene network in the nontumorigenic phenotype (EPN) with respect to the malignant ones (PC3, DU145 and LNCaP). Furthermore, the HOX gene modifications induced by cAMP make it possible to distinguish prostate metastatic cell lines according to the presence of andogen-receptors, active in LNCaP and absent in PC3 and DU145.[46]

In our cell system, exposure to cAMP mainly alters the expression of HOX genes situated on the HOX D locus and localized on chromosome 2q31-32. During embryonal development, Hox D genes play an important role in limb and digit generation.[47] In this same chromosomal area of the genome, a global control region has recently been identified upstream from the Hox D cluster, harbouring cis-regulatory DNA elements able to coordinate the expression not only of Hox D genes and their immediate surroundings, but also of phylogenetically unrelated genes lying several hundred kilobases from one another.[48] Besides the Hox D13 gene involved in prostate morphogenic defects in mutant mice,[49] this chromosomal area houses the genes CREB1 and CREB2 (ATF2) cyclic AMP activating transcription factors as well as the cAMP sensor guanine nucleotide exchange factor II gene (cAMP-GEFII), specifically active in the developing brain and regulating exocytosis in secretory cells. Furthermore, this same chromosomal area also houses the prostate-specific androgen-regulated gene PCGEM1, whose expression increases in high-risk prostate cancer patients.[50] PCGEM1 is a non coding RNA gene belonging to the microRNAs (miRNAs) family, involved in a number of important cell and developmental pathways (see conclusions). The chromosomal region 2q31-33 houses, in physical contiguity with PCGEM1, the genes neurexin-1 and NeuroD1. Neuronexins are polymorphic cell surface proteins coded by three genes (neuroxin 1, 2 and 3) expressed mostly in the brain and playing a role in cell-cell interaction.[51] NeuroD1 belongs to the "atonal" family of transcription factors, responsible for regulating the generation of a protosensory organ in Drosophila.[52] Thus, second messenger cAMP acts on a genomic region, molecularly coordinated with respect to epithelial-neuroendocrine conversion as it occurs along the tumour progression of prostate adenocarcinomas, presumably connected to androgen dependency and prostate phenotype through PCGEM1 and locus D HOX genes. It has recently been reported that the chromosomal area upstream of the HOX D locus contains nonrepetitive non coding regions extremely conserved during 500 million years of evolution, which may constitute an important insight into the evolution of mechanisms regulating associated gene complexes.[53] Coupling nuclear architecture and gene activity, spatial positioning in the genome is a functionally highly relevant, physiological and global phenomenon[54-55] (see conclusions).

We have selected one of these neurogenic genes located on the chromosomal area 2q31-33, Neuro D1, and tested the expression of its protein product along prostate cancer progression. The choice of Neuro D1 is supported by the observation that ectopic expression of Neuro D1 in Xenopus embryos causes conversion of epithelial cells into neurons.[56] Interestingly, NeuroD1 is actively expressed in vivo in advanced prostate cancers displaying a terminal stage of tumour differentiation as determined by an increased combined Gleason score.

These observations demonstrate HOX C6 as a crucial gene able to predict the clinical outcome of prostate cancers and whose expression is systematically altered in human bladder transitional cell carcinomas. Furthermore they suggest the in vivo involvement of locus D HOX genes together with a neurogenic gene from the 2q31-33 chromosomal area (Neuro D1) in the neuroendocrine differentiation characterizing tumour progression of human prostate cancers. Moreover they support the possibility to identify crucial unknown cancer genes through the study of the HOX gene network along tumour progression.

Kidney Cancer

The metanephric kidney architecture is created between the 5th and 15th weeks of human development. From the 16th to the 23rd weeks, following the filling in of the sulci between the lobes, the metanephros increases its growth, without completing it or acquiring kidney functionality.[57] Around the 23rd week of development, the foetal kidney starts to produce urine and continues to grow and acquire functions until birth. This process will be definitively completed during the first years of life.[58]

The molecular mechanisms involved in early kidney organogenesis require the expression, in the matanephric blastema, of paralogous group 11 Hox genes, Hox A11 and Hox D11,[59] to induce the outgrowth of ureteric bud from the Wolffian duct through the expression of several transcription factors (Wt1, Pax 2, Sall1, Fox C1 and Eya1),[60-64] of the nuclear protein forming,[65] the growth factors GDNF[66] and its tyrosine kinase receptor Ret[67] and coreceptor Gfrα1.[68] Wnt signals are crucial inductors of kidney tubologenesis, as deduced by the Wnt4 mutant mice lacking epithelial-mesenchymal transition and tubulogenesis.[69]

The crucial role played by Hox genes in the epithelial-mesechymal interaction during early kidney development is well established, with the expression of two of the three paralogous group 11 Hox genes, Hox A11 and Hox D11. Although single homozygous null mutation of either gene gives normal kidney, double mutants manifest hypoplasia of the kidneys[59] and removal of the last Hox 11 paralogous gene, Hox C11, results in the complete loss of metanephric kidney induction.[70] Little is known on the involvement of paralogous group 11 Hox genes on late kidney organogenesis. Other lumbo-sacral Hox genes (Hox A13 and Hox D13) are involved during early morphogenesis of the terminal part of the gut and urogenital tract.[71]

In order to verify the HOX network involvement in late human kidney organogenesis and to compare the HOX gene expression in kidney organogenesis and cancerogenesis we have recently studied the expression of the whole HOX gene network in foetal kidneys, from the 15th to the 38th week of human development, in pairs of normal-clear cell RCCs and in isolated clear cell RCCs (Cantile Personal Communication).

A comparison of the HOX network expression patterns between foetal, adult normal and tumorous kidney identifies, in the locus A, a decrease in the thoracocervical region of the network relative to the RCCs biopsies. Lumbo-sacral HOX D genes are all silent in foetal kidneys from the 15th to the 22nd week of development. HOX D9 and HOX D11 start to be expressed from the 23rd week, and remain active until the 38th developmental week. The expression of these genes can be temporally connected to kidney functions as the kidney starts functioning, for instance to produce urine, from the 23rd week. HOX D9 and HOX D11 are constitutively expressed in normal adult human kidneys and are the only lumbo-sacral HOX D genes active in primary epithelial tubular kidney cells to prove that the function of these homeoproteins is performed in tubular epithelial kidney cells. HOX D9 and HOX D11 are inactive in 76% and 100%, respectively, of the clear cell RCCs tested.

It has been suggested that, during tumour evolution, the gene profiles responsible for identifying specific cell phenotypes, undergo a de-differentiation programme towards early developmental stages. The patterns of lumbo-sacral HOX D gene expression in foetal kidneys (from the 15th to the 22nd week of development) overlap with the same patterns in clear cell RCCs suggesting that, in clear cell carcinomas of the kidney, the expression of lumbo-sacral HOX D genes marks a molecular de-differentiation process towards embryonic life.

During early metanephric kidney induction, paralogous group 11 HOX genes (HOX A11 and HOX D11) are able to induce the outgrowth of the ureteric bud from the Wolffian ducts, becoming inactive afterwards. Paralogous group 11 Hox genes triple mutant mice display bilateral kidney agenesis due to alterations in the epithelial-mesenchymal interaction, consequently preventing ureteric bud outgrowth. According to our data, HOX A11, whose effector is α8 integrin,[72] is again active at the 15th week of development, HOX C11 is expressed from the 18th week, HOX D11 starts again to be active from the 23rd week until birth and is always active in normal adult human kidneys. Thus, in timing, the expression of HOX D11 parallels the acquisition of kidney functions. The lack of expression we have detected in clear cell carcinomas of the kidney suggest the alteration of a kidney function related to HOX D11 homeoprotein. This function is probably connected to the role played by HOX D11 as well as the other lumbo-sacral HOX D genes in controlling epithelial-mesenchymal interaction in kidney as well as in other human organs.

Locus D HOX genes are localized on the chromosomal region 2q31-33. As we have previously reported the genes located on this genomic area are involved with epithelial-neurondocrine differentiation of human advanced prostate cancers.[44] This same chromosomal area of the genome houses the gene DIRC1, disrupted in renal carcinoma 1. Coupling nuclear architecture and gene activity, spatial positioning in the genome is a functionally highly relevant, physiological and global phenomenon.[55] The key involvement of HOX D11 in kidney cancers suggest the possibility to use the HOX D11 homeoprotein for diagnostic and therapeutic purposes in the clinical practice of clear cell kidney cancers.

Conclusions

When, fifteen years ago during an international cancer meeting, I presented a poster which linked HOX genes to human solid malignancies for the first time, one of the fathers of tumour biology asked what HOX genes had to do with human cancers. Since then, I have had to wait more than a decade to see the publication of an article describing "HOX genes as emerging stars in cancer research".[9]

The work we have carried out in the intervening years has allowed us to consider cancer as an anomalous structure growing inside the human body and following, from a cellular and architectural viewpoint, the rules controlling body shape during embryonic development. In consequence of this way of thinking we have, on the one hand, proposed the involvement of the whole HOX gene network in controlling phenotype cell identity and tridimensionality of tissues and organs and, on the other, described the implication of specific HOX genes or group of genes in the neoplastic alterations of particular cell phenotypes in specific organs. All this, despite the little understanding of the mechanisms involved, has already succeeded in the identification of specific HOX genes perturbed in certain types of human cancers with higher benefit for cancer patients as compared to well known oncogenes (HOX B13 versus ERBB2 in breast cancers—see breast cancer section).

In spite of this success, understanding the role of the HOX network in controlling phenotype cell identity and identifying HOX target genes are as yet unachieved objectives. Concerning the HOX target genes, we know for sure that they will be related to the majority of crucial eukaryotic cell functions.

I will end the chapter by describing three research areas potentially able to produce, in the near future, important achievements to increase our understanding of: (i) the function of the HOX gene network; (ii) the identification of the HOX target genes; (iii) the molecular basis of human cancers.

Transcriptional coactivators such as GCN5, p300/CBP and MOF represent the most intriguing of all the potential cofactors that might interact with HOX proteins to increase access to target genes on DNA. Initially, CBP was reported to interact specifically with HOX B7[73] as well as with HOX D4,[74] successively it was established that each of 14 HOX proteins from 11 separate HOX paralogous groups, bind to CBP and p300. CBP and p300 act on the general transcriptional machinery through the activity of a histone acetyltransferase (HAT) domain within the nucleosome core, facilitating and/or stabilizing access to target genes. Another histone modification, H3 K4 methylation, is connected with transcriptional activation in eukaryotic systems, Histone acetylation and H3 K4 methylation appear to interact with each other as demonstrated between Tritorax (TRX) and CBP in Drosophila[75] and MLL1 and CBP in human cells.[76] It has thus been proposed that HOX genes, instead of acting as transcriptional regulators, modulate acetyltransferase (HAT) and methyltransferase activities of coactivators.[77] Recently, the interaction of HOX B6 with CBP has been reported as able to repress the expression of the globin gene.[78]

It is generally accepted that the origin of the Hox gene network is due to duplications of a unique ancestral gene cluster during early vertebrate evolution. Recent indications suggest the evolution of several gene families located near the Hox clusters, parallel and contiguous to the Hox network. Many chromosome regions in the human genome exist in four similar copies, suggesting that the entire genome was duplicated twice in early vertebrate evolution, a concept called the 2R hypothesis.[79] The coordinate transcriptional regulation of an extended chromosomal region near the Hox D locus (see prostate cancer) and its extreme conservation during 500 million years of evolution constitutes a proof of the positional connections inside the genome.[48,53] Thus, chromosome and genes are non randomly positioned in the cell nucleus. Recent indications support the view of a function connected to gene activity and genome stability. The mechanism and significance of positioning are unknown. What molecular mechanisms might be responsible for the positioning of gene and chromosome? Could physical contiguity be a factor in common metabolic processes? Might there be changes in gene positioning consequent to physiological and/or pathological situations?[55] The lack of answers to these questions is attributable to the single-gene analyses so far performed. The post-genomic era with its high-throughput analysis of large sets of genes coupled to bioinformatics will end up by identifying of a new dimension due to spatial positioning on the genome functioning.

MicroRNAs are a class of molecules, 21 to 25 nucleotides in length, that are able to negatively regulate gene expression. These small RNAs are transcribed in the nucleus (pri-miRNA) before being exported to the cytoplasm where they mature through the enzyme Dicer in RNA-induced silencing complex (RISC). Base pairing between miRNA and its target induce RISC to destroy mRNA or to block its translation into protein. Several hundred miRNAs have been identified in animals and plants. The expression profiling of 217 miRNAs in multiple human tumor types allows the hierarchical clustering of the samples according to the developmental origin of tissues, to diversify normal from cancerous tissues and to associate miRNA expression to cell differentiation.[80] Previously, tissue specific expression of miRNAs has been reported during embryogenesis. Two of these miRNAs, miR-196 and miR-10b are localised inside the Hox network: miR 196 is found in the Hox B8 gene, and interacts with the other paralogous group 8 Hox genes as well as with Hox A7,[81] miR-10b is located in the Hox D locus between Hox D4 and Hox D8.[82] Other miRNAs are located in contiguity to the HOX network (see PCGEM and prostate cancer). It has recently been suggested that miRNAs act in fine-tuning specific Hox mRNA expression patterns during mouse development.[83] The possibility to characterize human solid tumours by the combined expression of the HOX gene network and miRNAs could succeed in the identification of a molecular atlas of human cancers.

Acknowledgements

I am grateful to my collaborators Monica Cantile, Luca Cindolo, Giulia Schiavo, Vincenzo D'Antò and Alfredo Procino whose enthusiasm has made it possible to carry out the work I have described in this chapter.

References

1. Scott MP. Hox genes, arms and the man. Nature 1997; 15:117-118.
2. Cillo C, Faiella A, Cantile M et al. Homeobox genes and cancer. Exp Cell Res 1999; 248(1):1-9.
3. Cillo C, Cantile M, Faiella A et al. Homeobox genes in normal and malignant cells. J Cell Physiol 2001; 188(2):161-169.
4. Garcia-Bellido A. Genetic control of wing disc development in Drosophila. Elsevier, Amsterdam: CIBA Foundation Symposium Edition, 1975:29.
5. Cillo C. HOX genes in human cancers. Invasion and Metastasis 1994-95; 14:38-49.
6. Castelli-Gair Hombria J, Lovergrove B. Beypnd homeosis- HOX function in morphogenesis and organogenesis. Differentiation 2003; 71:461-476.
7. Sinisi AA, Chieffi P, Pasquali D et al. EPN: A novel epithelial cell line derived from human prostate tissue. In Vitro Cell Dev Biol Anim 2002; 38:165-172.
8. Leroy P, Berto F, Bourget I et al. Downregulation of Hox A7 is required for cell adhesion and migration on fibronectin during early HL-60 monocytic differentiation. J Leukoc Biol 2004; 75:680-8.
9. Chen H, Sukumar S. HOX genes: Emerging stars in cancer. Cancer Biol Ther 2003; 2:524-5.
10. Abate-Shen C. Deregulated homeobox gene expression in cancer: Cause or consequence? Nat Rev Cancer 2002; 2:777-85.

11. Friedmann Y, Daniel CA, Strickland P et al. Hox genes in normal and neoplastic mouse mammary gland. Cancer Res 1994; 54(22):5981-5985.
12. Chariot A, Moreau L, Senterre G et al. Retinoic acid induces three newly cloned HOXA1 transcripts in MCF7 breast cancer cells. Biochem Biophys Res Commun 1995; 215(2):713-720.
13. Chariot A, Castronovo V. Detection of HOXA1 expression in human breast cancer. Biochem Biophys Res Commun 1996; 222(2):292-297.
14. Boylan JF, Lohnes D, Taneja R et al. Loss of retinoic acid receptor gamma function in F9 cells by gene disruption results in aberrant Hoxa-1 expression and differentiation upon retinoic acid treatment. Proc Natl Acad Sci USA 1993; 90(20):9601-9605.
15. Chariot A, Castronovo V, Le P et al. Cloning and expression of a new HOXC6 transcript encoding a repressing protein. Biochem J 1996; 319(Pt 1):91-97.
16. Boudreau N, Andrews C, Srebrow A et al. Induction of the angiogenic phenotype by Hox D3. J Cell Biol 1997; 139(1):257-264.
17. Care A, Felicetti F, Meccia E et al. HOXB7: A key factor for tumor-associated angiogenic switch. Cancer Res 2001; 61(17):6532-6539.
18. Raman V, Martensen SA, Reisman D et al. Compromised HOXA5 function can limit p53 expression in human breast tumours. Nature 2000; 405(6789):974-978.
19. Raman V, Tamori A, Vali M et al. HOXA5 regulates expression of the progesterone receptor. J Biol Chem 2000; 275(34):26551-26555.
20. Chen F, Capecchi MR. Paralogous mouse Hox genes, Hoxa9, Hoxb9, and Hoxd9, function together to control development of the mammary gland in response to pregnancy. Proc Natl Acad Sci USA 1999; 96(2):541-546.
21. Cantile M, Pettinato G, Procino A et al. In vivo expression of the whole HOX gene network in human breast cancer. Eur J Cancer 2003; 39:257-64.
22. Ma XJ, Wang Z, Ryan PD et al. A two-gene expression ratio predicts clinical outcome in breast cancer patients treated with tamoxifen. Cancer Cell 2004; 5:607-16.
23. Singletary SE, Allred C, Ashley P et al. Revision of the american joint committee on cancer staging system for breast cancer. J Clin Oncol 2002; 20:3628-36.
24. Dalton LW, Pinder SE, Elston CE et al. Histologic grading of breast cancer: Linkage of patient outcome with level of pathologist agreement. Mod Pathol 2000; 13:730-5.
25. Bonner RF, Emmert-Buck M, Cole K et al. Laser capture microdissection: Molecular analysis of tissue. Science 1997; 278(5342):1481-1483.
26. Sandberg AA. Cytogenetics and molecular genetics of bone and soft-tissue tumors. Am J Med Genet 2002; 115:173-82.
27. Cordon-Cardo C. Molecular alterations in bladder cancer. Cancer Surv 1998; 32:115-31.
28. Fujimoto K, Yamada Y, Okajima E et al. Cancer Res 1992; 52:1393-8.
29. Wada T, Louhelainen J, Hemminki K et al. Clin Cancer Res 2000; 6:610-5.
30. Shinohara N, Koyanagi T. Ras signal transduction in carcinogenesis and progression of bladder cancer: Molecular target for treatment? Urol Res 2002; 30:273-81.
31. Adshead JM, Ogden CW, Penny MA et al. The expression of PAX5 in human transitional cell carcinoma of the bladder: Relationship with de-differentiation. BJU Int 1999; 83:1039-44.
32. Cantile M, Cindolo L, Napodano G et al. Hyperexpression of locus C genes in the HOX network is strongly associated in vivo with human bladder transitional cell carcinomas. Oncogene 2003; 22:6462-8.
33. Simon R, Struckmann K, Schraml P et al. Amplification pattern of 12q13-q15 genes (MDM2, CDK4, GLI) in urinary bladder cancer. Oncogene 2002; 21:2476-83.
34. Yoon SJ, LeBlanc-Straceski J, Ward D et al. Organization of the human keratin type II gene cluster at 12q13. Genomics 1994; 24:502-8.
35. Jave-Suarez LF, Winter H, Langbein L et al. HOXC13 is involved in the regulation of human hair keratin gene expression. J Biol Chem 2002; 277:3718-26.
36. Lussier M, Filion M, Compton JG et al. The mouse keratin 19-encoding gene: Sequence, structure and chromosomal assign. Gene 1990; 95:203-13.
37. Moll R, Achtstatter T, Becht E et al. Cytokeratins in normal and malignant transitional epithelium. Maintenance of expression of urothelial differentiation features in transitional cell carcinomas and bladder carcinoma cell culture lines. Am J Pathol 1988; 132:123-44.
38. Spitz F, Gonzalez F, Duboule D. A global control region defines a chromosomal regulatory landscape containing the HoxD cluster. Cell 2003; 113:405-417.
39. Hostikka SL, Capecchi MR. The mouse Hoxc11 gene: Genomic structure and expression pattern. Mech Dev 1998; 70:133-45.
40. Gleason DF. Histologic grading and prostate cancer: A perpective. Hum Pathol 1992; 23:273-279.

41. Roberts SG, Blute ML, Bergstralh EJ et al. PSA doubling time as a predictor of clinical progression after biochemical failure following radical prostatectomy for prostate cancer. Mayo Clin Proc 2001; 76:576-81.
42. Singh D, Febbo PG, Ross K et al. Gene expression correlates of clinical prostate cancer behavior. Cancer Cell 2002; 1:203-9.
43. Miller GJ, Miller HL, van Bokhoven A et al. Aberrant HOXC expression accompanies the malignant phenotype in human prostate. Cancer Res 2003; 63:5879-5888.
44. Cantile M, Kisslinger A, Cindolo L et al. cAMP induced modifications of HOX D gene expression in prostate cells allow the identification of a chromosomal area involved in vivo with neuroendocrine differentiation of human advanced prostate cncers. J Cell Physiol 2005; 205:202-210.
45. Bang YJ, Pirnia F, Fang WG et al. Terminal neuroendocrine differentiation of human prostate carcinoma cells in response to increased intracellular cyclic AMP. Proc Natl Acad Sci USA 1994; 91:5330-5334.
46. Lin DL, Whitney MC, Yao Z et al. Interleukin-6 induces androgen responsiveness in prostate cancer cells through upregulation of androgen receptor expression. Clin Cancer Res 2001; 7:1773-1781.
47. Zakany J, Kmita M, Duboule D. A dual role for Hox genes in limb anterior-posterior asymmetry. Science 2004; 304:1669-1672.
48. Spitz F, Gonzalez F, Duboule D. A global control region defines a chromosomal regulatory landscape containing the HoxD cluster. Cell 2003; 113:405-417.
49. Podlasek CA, Duboule D, Bushman W. Male accessory sex organ morphogenesis is altered by loss of function of Hoxd-13. Dev Dyn 1997; 208:454-465.
50. Petrovics G, Zhang W, Makarem M et al. Elevated expression of PCGEM1, a prostate-specific gene with cell growth-promoting function, is associated with high-risk prostate cancer patients. Oncogene 2004; 23:605-611.
51. Kleiderlein JJ, Nisson PE, Jessee J et al. CCG repeats in cDNAs from human brain. Hum Genet 1998; 103:666-673.
52. Niwa N, Hiromi Y, Okabe M. A conserved developmental program for sensory organ formation in Drosophila melanogaster. Nat Gen 2004; 36:293-297.
53. Sabarinadh C, Subramanian S, Tripathi A et al. Extreme conservation of noncoding DNA near HoxD complex of vertebrates. BMC Genomics 2004; 5(1):75.
54. Casolari JM, Brown CR, Komili S et al. Genome-wide localization of the nuclear transport machinery couples transcriptional status and nuclear organization. Cell 2004; 117:427-39.
55. Misteli T. Spatial positioning; a new dimension in genome function. Cell 2004; 119:153-6.
56. Lee JE, Hollenberg SM, Snider L et al. Conversion of Xenopus ectoderm into neurons by NeuroD, a basic helix-loop-helix protein. Science 1995; 268:836-844.
57. Saxèn L. Organogenesis of the kidney. Cambridge, UK: Cambridge Univ. Press, 1987:1-173.
58. Potter EL. Normal and abnormal development of the kidney. Chicago: Year Book Medical publishers Inc, 1972.
59. Patterson LT, Pembaur M, Potter SS. Hoxa11 and Hoxd11 regulate branching morphogenesis of ureteric bud in the developing kidney. Development 2001; 128:2153-2161.
60. Kreidberg JA, Sariola H, Loring JM et al. Wt1 is required for early kidney development. Cell 1993; 74:679-691.
61. Torres M, Gomez-Pardo E, Dressler GR et al. Pax2 controls multiple steps of urogenital development. Development 1995; 121:4057-4065.
62. Nishinakamura R, Matsumoto Y, Nakao K et al. Murine homolog of SALL1 is essential for ureteric bud invasion in kidney development. Development 2001; 128:3105-3115.
63. Kume T, Deng K, Hogan BL. Murine forkhead/winged helix genes Foxc1 (Mf1) and Foxc2 (Mfh1) are required for the early organogenesis of the kidney and urinary tract. Development 2000; 127:1387-1395.
64. Xu PX, Adams J, Peters H et al. Eya1-deficient mice lack ears and kidneys and show abnormal apoptosis oforgan primordia. Nat Genet 1999; 23:113-117.
65. Davies JA, Fisher CE. Genes and proteins in renal development. Exp Nephrol 2002; 10:102-113.
66. Pichel JG, Shen L, Sheng HZ et al. Defects in enteric innervation and kidney development in mice lacking GDNF. Nature 1996; 382:73-76.
67. Schuchardt A, D'Agati V, Larsson-Blomberg L et al. Defects in the kidney and enteric nervous system of mice lacking the tyrosine kinase receptor Ret. Nature 1994; 367:380-383.
68. Cacalano G, Farinas I, Wang LC et al. GFRalpha1 is an essencial receptor component for GDNF in the developing nervous system and kidney. Neuron 1998; 21:53-62.
69. Stark K, Vaino S, Vassileva G et al. Epithelial transformation of metanephric mesenchyme in the developing kidney regulated by Wnt-4. Nature 1994; 372:679-683.

70. Wellik DM, Hawkes PJ, Capecchi MR. Hox11 paralogous genes are essencial for metanephric kidney induction. Genes and Development 2002; 16:1423-1432.
71. Warot X, Fromental-Ramain C, Fraulob V et al. Gene dosage-dependent effects of the Hox a13 and Hox d13 mutations on morphogenesis of the terminal parts of the digestive and urogenital tracts. Development 1997; 124:4781-4791.
72. Valerius MT, Patterson LT, Yuxin Feng et al. Hox a11 is upstream of Integrinα8 expression in the developing kidney. PNAS 2002; 99:8090-8095.
73. Chariot A, van Lint C, Chapelier M et al. CBP and histone deacetylase inhibition enhance the transactivation potential of the HOXB7 homeodomain-containing protein. Oncogene 1999; 18:4007-14.
74. Saleh MI, Koh YM, Tan SC et al. Clean-up, detection and determination of salbutamol in human urine and serum. Analyst 2000; 125:1569-72.
75. Petruk S, Sedkov Y, Smith S et al. Trithorax and dCBP acting in a complex to maintain expression of a homeotic gene. Science 2001; 294:1331-4.
76. Ernst P, Wang J, Huang M et al. MLL and CREB bind cooperatively to the nuclear coactivator CREB-binding protein. Mol Cell Biol 2001; 21:2249-58.
77. Shen WF, Krishnan K, Lawrence HJ et al. The HOX homeodomain proteins block CBP histone acetyltransferase activity. Mol Cell Biol 2001; 21:7509-22.
78. Shen WF, Chrobak D, Krishnan K et al. HOX B6 protein is bound to CREB-binding protein and represses globin expression in a DNA binding-dependent, PBX interaction-independent process. The Jour Biol Chem 2004; 279:39895-39904.
79. Larhammar D, Lundin LG, Hallbook F. The human Hox-bearing chromosome regions did arise by block or chromosome (or even genome) duplications. Genome Res 2002; 12:1910-20.
80. Lu J, Getz G, Miska EA et al. MicroRNA expression profiles classify human cancers. Nature 2005; 435:834-8.
81. Yekta S, Shih IH, Bartel DP. MicroRNA-directed cleavage of HOXB8 mRNA. Science 2004; 304:594-6.
82. Mansfield JH, Harfe BD, Nissen R et al. MicroRNA-responsive 'sensor' transgenes uncover Hox-like and other developmentally regulated patterns of vertebrate microRNA expression. Nat Genet 2004; 36:1079-83.
83. Cobb J, Duboule D. Tracing microRNA patterns in mice. Nat Genet 2004; 36:1033-4.

Appendix

Hypothesis:
Pulling Forces Acting on *Hox* Gene Clusters Cause Expression Collinearity[a]

Spyros Papageorgiou*

Abstract

The development of normal patterns along the primary and secondary vertebrate axes depends on the regularity of early *Hox* gene expression. During initial stages, these expression events form a sequential pattern of partially overlapping domains along the anteroposterior axis in coincidence with the 3' to 5' order of the genes in the *Hox* cluster (spatial collinearity). In addition, the genes are activated one after the other in the 3' to 5'order (temporal collinearity). These features are poorly understood within the framework of Molecular Genetics. A model was proposed according to which physical forces act on *Hox* clusters as a result of signaling from morphogen gradients. The model can explain the collinearity of *Hox* gene expression along the primary and secondary body axes. The increase of the morphogen concentration is accordingly followed by an increase of the force acting on the cluster. The genes are sequentially translocated, in the 3' to 5' order, toward the interchromosome domain where they are exposed to transcription factors for activation. The above geometrodynamic approach reproduces most collinearity data. Recent experiments verify the above prediction of sequential 3' to 5' *Hox* gene translocations in the interchromosome domain. Furthermore, it seems that these translocations, combined with cluster decondensations, are caused by attractive forces acting on the 3' end of the cluster and pulling the genes out of the chromosome territory. Additional experiments are proposed in order to specify the origin of the forces.

Introduction

In 1978, after a long series of classical genetic studies in *Drosophila*, E. B. Lewis established a correlation between the activation pattern of the genes of the bithorax complex (*BX-C*) along the antero-posterior axis of the embryo and the proximo-distal location of these genes along the chromosome.[1] This astonishing correlation (coined collinearity) proved to be a property extending to orthologous genes of all metameric animals, humans included.[2] It turns out that these genes contain a conserved sequence of 180 bp, the so-called homeobox, which was first discovered in the *Antennapedia* and the other homeotic genes of *Drosophila*.[2,3]

---

[a]This paper appeared in the Special Issue 'Developmental Morphodynamics' of the International Journal of Developmental Biology, 2006; 50:301-308.

*Corresponding Author: Spyros Papageorgiou—Institute of Biology, NRC 'Demokritos', 153 10 Aghia Paraskevi, Athens, Greece. Email: spapage@bio.demokritos.gr

HOX Gene Expression, edited by Spyros Papageorgiou. ©2007 Landes Bioscience and Springer Science+Business Media.

Compared to *Antennapedia*, numerous homologue genes were found in many other genomes. These were named *Hox* genes and it turns out that they are grouped in complexes called *HOX* clusters. As in the *Drosophila*, *Hox* gene mutations cause severe homeotic transformations and malformations of the embryonic body plan. Because of their importance in axial patterning, several homologue clusters have been formed by duplication in the course of evolution and their genes play significant and complementary roles in development. While *Drosophila* has only one homeotic complex (*HOM-C*), vertebrates have four such paralogous clusters *Hoxa*, *Hoxb*, *Hoxc* and *Hoxd* each one located in a different chromosome.[3] Every paralogous cluster has a variable number of genes (9 to 11) numbered from 1 to 13 in their physical order along the 3' to 5' direction on the chromosome. (In every cluster some genes of the above numbering are missing).

It was soon realized that the anteroposterior boundaries of *Hox* gene expression along the axis of the mouse embryo followed a collinear relationship similar to the *BX-C* correlation established by Lewis in *Drosophila*.[4,5] Namely, the anterior boundary of every expression was shifted along the posterior direction following the order 1, 2, 3,...of the genes in the cluster.[3] This was characterized as *spatial collinearity* for the primary axis whereas a modified spatial collinearity was also observed for the mouse and chick limb buds[6,7] (Fig. 1). In this case, spatial collinearity takes the following form at the initial stages of activation: for the *Hoxa* genes at the 5' end of the cluster, the expression domains create a nested pattern of partially overlapping regions along the proximo-distal axis of the bud. The expression domain of the last gene, *Hoxa13*, is limited at the distal tip of the bud. A similar pattern of nested expression domains is also observed for the 5' genes of *Hoxd* cluster with the last gene, *Hoxd13*, being expressed at the posterior-distal boundary of the bud.[8]

Another interesting observation in vertebrate development concerns the time dependence of *Hox* gene expression: the initiation of expression follows the physical order of the genes in the cluster. Gene1 is expressed first followed by gene2 etc. This sequence in time of gene activation is termed *temporal collinearity*.[9] A third kind of *Hox* collinearity results from the fact that the intensity at the anterior part of an expression domain is strong compared to the posterior expression region.[2] In particular and in contrast to the anterior boundary, the posterior expression boundary is faint and unclear. Therefore, at every location the expression of a posterior gene is stronger and dominates over any other overlapping anterior gene (Fig. 1). This property can be related to the *posterior prevalence* if we assume that the protein dominance reflects the dominance in strength of its gene expression.[2] One can think of an equivalence between *posterior prevalence* and *quantitative collinearity* observed for the 5' gene expressions of the *Hoxd* cluster in the limb bud:[10,11] it turns out that distally in the bud the expression intensity systematically increases following the order *Hoxd10,...,Hoxd13*.

The above features of *Hox* gene collinearity have been meticulously analyzed for more than 25 years in many organisms ranging from *Drosophila* to humans. Many attempts have been made to justify these surprising regularities but, up to now, no convincing solution has been possible based on molecular genetics and biochemical processes. Furthermore, the changes of a gene location in the cluster are associated with systematic alterations of its expression mode.[11] This leads to the hypothesis that the chromatin deforming forces in the cluster may be responsible for the expression pattern of *Hox* genes. I think therefore that the mechanistic nature of collinearity phenomena requires a geometrodynamical approach. I have put forward a broad framework of physical principles combined with well founded biological facts which can reasonably reproduce the observed collinearity data.[12,13] The guiding dogma in this pursuit is that what happens is what can happen: phenomena that have been observed in other occasions are adopted and applied in the present formulation as long as they support the central hypothesis. This is legitimate since, in this way, the constructed model is compatible with well established facts and principles. The details of the mechanisms involved can only be determined and confirmed by experiment. It is instructive to give an example of how this has to be done. The model, in its broad formulation, proposes that collinearity is the result of

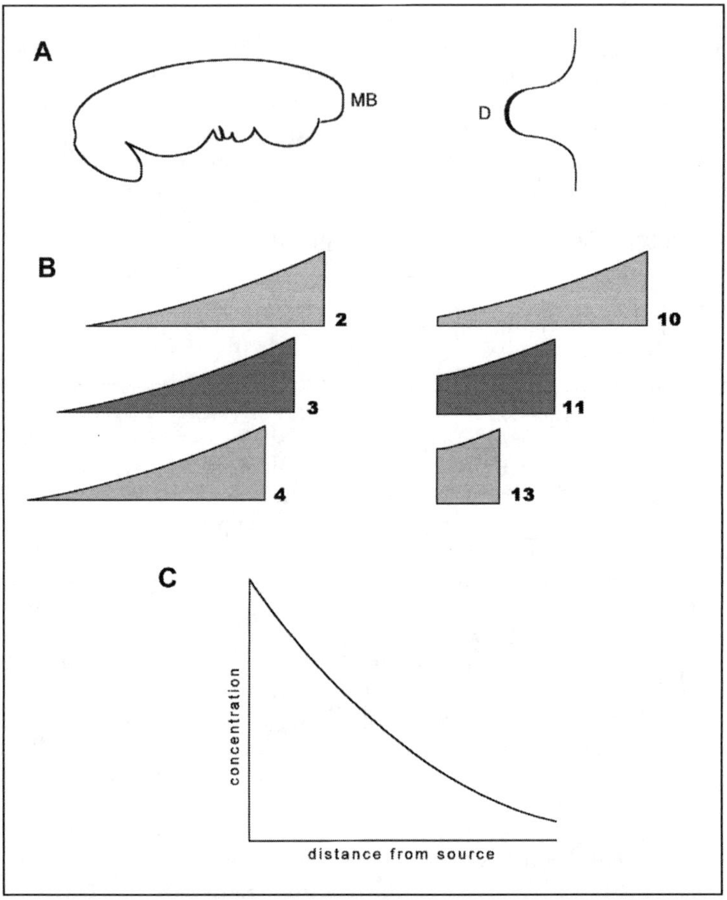

Figure 1. Diagrams of embryos and morphogen gradients. A) On the left, a 12 day mouse embryo is represented with the midbrain (MB) to the right boundary of the embryo axis and the posterior end to the left. On the right, the limb bud is represented with the distal tip (D) to the left. B) On the left, the schematic expression domains of *Hoxb2, Hoxb3* and *Hoxb4* are shown along the primary axis of the embryo. Expression is stronger anteriorly (right side) and the expression domains are shifted along the anterior-posterior axis following the order 2, 3, 4 (posterior prevalence). On the right, the expression domains of *Hoxa10, Hoxa11* and *Hoxa13* in the limb bud at the initial stages (around 24) are depicted. The domains form a nested pattern while distally the intensity of expression increases in the order *Hoxa10, Hoxa11, Hoxa13* (quantitative collinearity). C) A morphogen gradient with a peak at the posterior end of the vertebrate embryo or the distal tip of the limb bud.

physical forces translocating the *Hox* cluster toward a specific region where gene activation can occur. The nature, direction and application point of the forces cannot be known a priori since several possibilities can produce the expected result. After the formulation of the model an experiment has confirmed these translocations during gene activation of the *Hoxb* cluster.[14] Furthermore, this experiment has indicated where and how the force should apply on the cluster. This is helpful since it allows to write down, in the following, a detailed and more concrete version of the model. A short description is given below.

A Physical Model for *Hox* Gene Collinearity

The model functions at two levels in space:

1. At a multicellular ('macroscopic') level, a morphogen gradient is established over several cell diameters in morphogenetic fields of linear size approaching 1mm[15,16] as shown in Figure 1 for the developing chick limb bud. For many years, persistent efforts aimed to explain how these gradients are created.[17] Some recent findings corroborate a mechanism of local morphogen production and subsequent extracellular spreading and degradation of the morphogen. The spreading is diffusion-based associated with or without endocytosis depending on the specific developing system.[18,19] Passive diffusion may be combined with secondary procedures like morphogen transport through membranes[20] or it may operate in parallel with processes like growth.[21] Such secondary effects might hide the basic features of diffusion and, in order to determine the actual signal propagating mechanism, it is useful to compare the observed data with the features of signal traveling and their dependence on time and source intensity as expected in the case of pure diffusion.[22,23] For simplicity here it is assumed that at the posterior end of the vertebrate embryo axis (or the distal end of the limb bud) a source starts producing a morphogen which is spread in the morphogenetic field through a mechanism based on diffusion, as described above. After a transient time interval of morphogen net production and degradation in every cell, a decreasing steady state distribution is reached of exponential form with its peak at the source area[22,24] (Fig. 1).
2. At the nuclear ('microscopic') level, it is assumed that the tightly packaged *Hox* clusters take the shape of an elongated rigid body whose length is of the order of 500 nm.[12] This is consistent with the findings that at interphase chromatin fibers are supercoiled into larger chromonema fibers whose diameters are about 80 nm[25,26] as shown in Figure 2. These tightly clustered *Hox* complexes, when inactive, they are embedded inside the chromosome territory (CT) with their regulatory regions inaccessible for transcription.[27,28] There is strong evidence that, with chromatin restructuring, gene activation occurs at the surface of the chromosome territories when genes enter in the interchromosome domain (ICD) where they can be reached by the transcription factors[29-31] (Figs. 3, 4). The CT forms a meandering dense structure from which decondensing chromatin fibers extend (Fig. 4).[29,32] The ICD consists of a network of channels around and through CTs. These channels are connected with nuclear pores at the nuclear surface.[32] Transcription factors (TF) activating *Hox* genes are confined in ICD together with other mobile regulatory molecules.[27] It is furthermore assumed that the density of TF decreases away from the chromosome surface. This agrees with the observation of a concentration gradient of transcription factor Stat1 imported in the nucleus via a carrier-free (diffusion) process.[33]

The morphogen-signal of the macroscopic level is transduced and uniformly amplified inside the cell so that the product molecules [PM] are produced and they act on the microscopic nuclear level. The transportation of SMAD2 inside the nucleus is an example of such a transduction and amplification.[34,35] Furthermore, it was recently observed that Smad regulation in the nucleus is achieved through a physical interaction between the inner nuclear membrane MAN1 and R-Smads.[36] The case of SMAD2 is not unique. Many other molecules are imported and deposited in specific locations inside the nucleus as for instance the DSH protein which is the product of Wnt signal transduction.[37] Thus, the morphogen concentration acting macroscopically is associated to a corresponding concentration of [PM] endowed with suitable physicochemical properties. For example, [PM] can be polar molecules that bind on the chromosome surface and collectively create an electric field.[12,13] The resulting Coulomb force (attractive or repulsive) acts then on the (negatively charged) *Hox* cluster (Fig. 4).

The case of electric repulsion pushing the cluster out of CT has been worked out and the existing data of spatial, temporal and quantitative collinearity have been well reproduced.[12,13] Furthermore, the results of genetic deletions and duplications and the associated modifications of *Hoxd* expressions in the mouse limb bud[11] were also well described.[13]

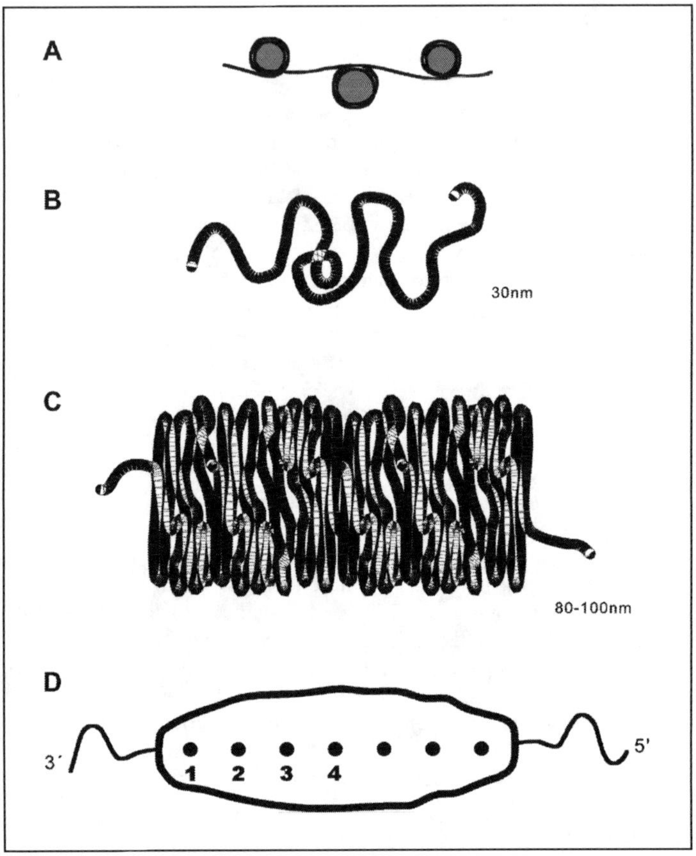

Figure 2. Hierarchical levels of DNA packaging: A) Nuclesomes (11 nm) arranged as beads-on-a-string. B) Folded chromatin fiber of 30 nm diameter. C) Condensed chromatin fiber (chromonema) of 100 nm diameter. D) A *Hox* cluster schematically shown as a rod whose length is about 500 nm.

Attractive Electric Forces

Chambeyron and Bickmore have induced gene transcription of the *Hoxb* cluster during the differentiation of murine ES cells.[14] When *Hoxb1* and *Hoxb9* are inactive they are located inside the CT. After induction with retinoic acid, the chromatin is decondensed and an extrusion of *Hoxb1* from the CT follows in association with *Hoxb1* expression while *Hoxb9* remains inside CT. Later *Hoxb9* is also shifted from inside the CT and it stays in the ICD but close to the CT surface. At the same time *Hoxb1* and the decondensed chromatin fiber performs a 'choreographed looping' in ICD with an overall movement toward the centre of the nucleus.[14] This behaviour of *Hoxb1* and *Hoxb9* is in agreement with the fiber translocation according to the model.

Instead of a repulsive force applied on the 5' end of the cluster as described above, the observation of Chambeyron and Bickmore could be understood as the result of an attractive force that acts on the 3' end of *Hoxb* cluster and pulls the fiber toward ICD. This situation, in principle, is similar to the experiment of Cui and Bustamante who measured chromatin extensions (decondensations) as the result of pulling forces for a wide range of variation of

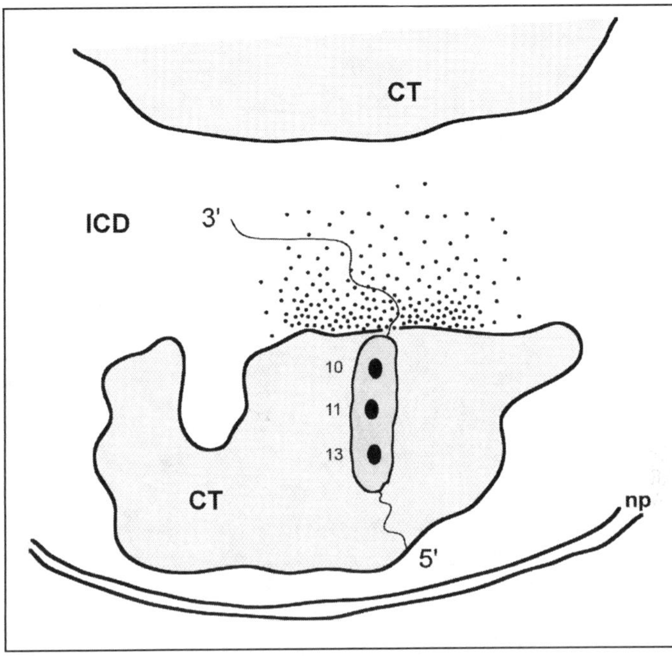

Figure 3. Schematic drawing of a nuclear section. The single Chromosome Territory (CT) is traversed by the Interchromosome Domain (ICD) which forms a set of channels around and through the chromosome.[32] Transcription factors (black spots) and other regulating molecules circulate in the ICD. Before activation the *Hox* cluster is hidden inside Chromosome Territory. At the bottom is the envelop of the nuclear periphery (np).

the stretch modulus up to 25 pN.[38] It is assumed therefore that, in response to the extracellular morphogen signals, positive polar molecules [PPM] are deposited opposite the 3' end of the cluster as shown in Figure 4. An appropriate electric field is then acting on the negatively charged *Hox* cluster. The resulting electric force pulls the chromatin fiber toward the ICD and the packaged cluster is decondensed with the chromatin tethered and stretched out of the CT. Note that the same physical principles govern electrophoresis in experiments of stretching or active transport of DNA molecules and other biopolymers.[39] Electrophoretic forces are widely used to measure the elastic properties and large-scale conformational changes of chromatin fibers.[40] The deformations of tethered DNA in an electric field are similar to the stretching deformations due to hydrodynamic flow[41,42] and it turns out that this chromatin remodeling is necessary for DNA transcription.[43]

In cells exposed to low morphogen concentrations the resulting electrophoretic forces are weak and the *Hox* cluster translocations are accordingly small (Fig. 4A). A sequential increase of the morphogen concentration will respectively produce an increasing dragging force and the *Hox* genes will be translocated into ICD following their physical order: 1, 2, 3... (Fig. 4B). It is assumed that the density of transcription factors decreases away from the CT surface and this generates quantitative collinearity (Fig. 4). According to Figure 4A proximally only *Hoxa 10* is expressed and the intensity of expression is strong since the gene is close to the CT surface. Distally, all genes are expressed with increasing intensity (*Hoxa 10*, *Hoxa 11*, *Hoxa 13*) because of their relative position in ICD (Fig. 4B) in agreement with quantitative collinearity of Figure 1B.

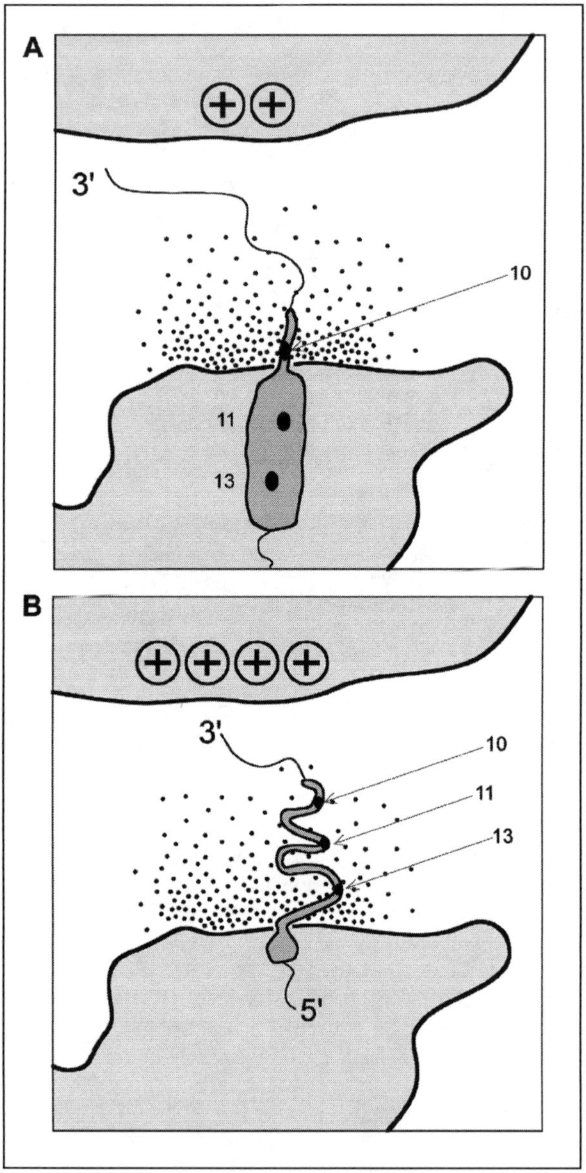

Figure 4. Detail of Figure 3 around the exit pore of *Hoxa* cluster. A) A small electrophoretic force from the opposite Chromosome Territory surface pulls *Hoxa10* out of the CT. B) A stronger force stretches the fiber and pulls all three *Hoxa* genes inside the Interchromosome Domain.

Correlation of Morphogen Thresholds and Gene Translocations

The present model can correlate morphogen thresholds with gene translocations and transcription. As an example, in Figure 5 is shown the normal morphogen gradient M which controls the activation of the 5' genes of *Hoxa* cluster in the limb bud. The horizontal lines L and H represent the lower and higher morphogen thresholds respectively for *Hoxa13* activation in the distal region of the bud. For M the expression domain of *Hoxa13* is (AP). When an

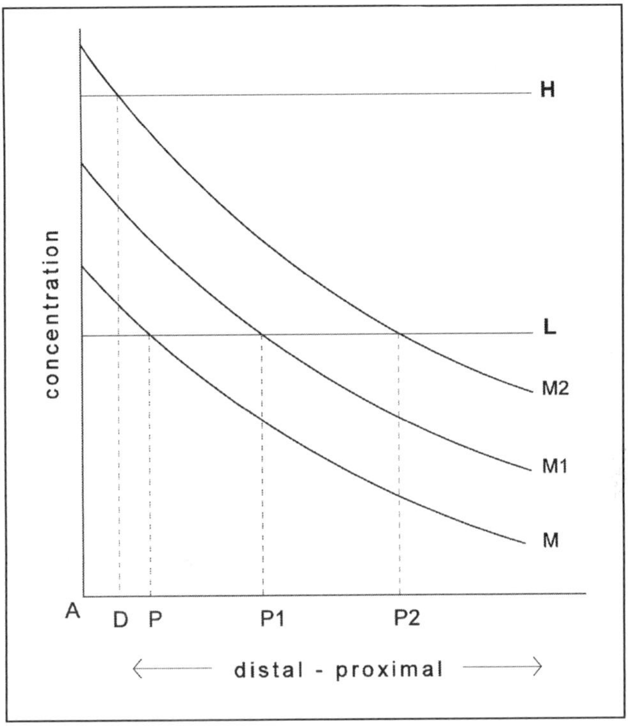

Figure 5. Morphogens and thresholds. Normal morphogen gradient M with L (lower) and H (higher) thresholds for *Hoxa13* activation. Additional morphogen sources at the tip increase M to M1 or M2.

additional morphogen source is inserted at the distal tip, the gradient increases to M1 and the *Hoxa13* activation domain expands to (AP1). The region (PP1) represents the domain of ectopic gene expression. If an even stronger morphogen source is inserted, the gradient increases further to M2 exceeding eventually the higher threshold H at the tip. In this case the ectopic expression reaches the proximal boundary P2 but in the region (AD), where the morphogen concentration is higher than H, the gene expression disappears.

It is instructive to interpret the above gene response to morphogen thresholds in terms of the geometry and dynamics of cluster translocations. For simplicity *Hoxa13* is isolated from the cluster and it is depicted in Figure 6 for the different stages of gene activation. When *Hoxa13* is hidden in the interior of CT (Fig. 6A) the gene is inactive and this case corresponds to proximal cells where the morphogen level is below L in Figure 5. When the morphogen concentration increases the associated electrophoretic force pulls the gene just above the chromosome surface (Fig. 6B). For the normal gradient M this occurs in proximal cells of the domain (AP) in Figure 5. When the morphogen concentration increases further, the gene is accordingly pulled stronger and it is shifted toward the border of the transcription factor region (Fig. 6C). This corresponds to the distal boundary D which is reached for the gradient M2 of Figure 5. The morphogen level there is just below H. Finally, for M2 the domain (AD) has morphogen concentrations higher than H, therefore the associated gene translocations are beyond the border of the transcription factor region and gene expression disappears (Fig. 6D). Such an expression attenuation has been observed for *Hoxa13* in the chick limb when a FGF4 bead was inserted at the distal tip of the bud.[23] However, in this case no proximal ectopic expression was observed (PP2 in Fig. 5), as would be expected according to the model. A

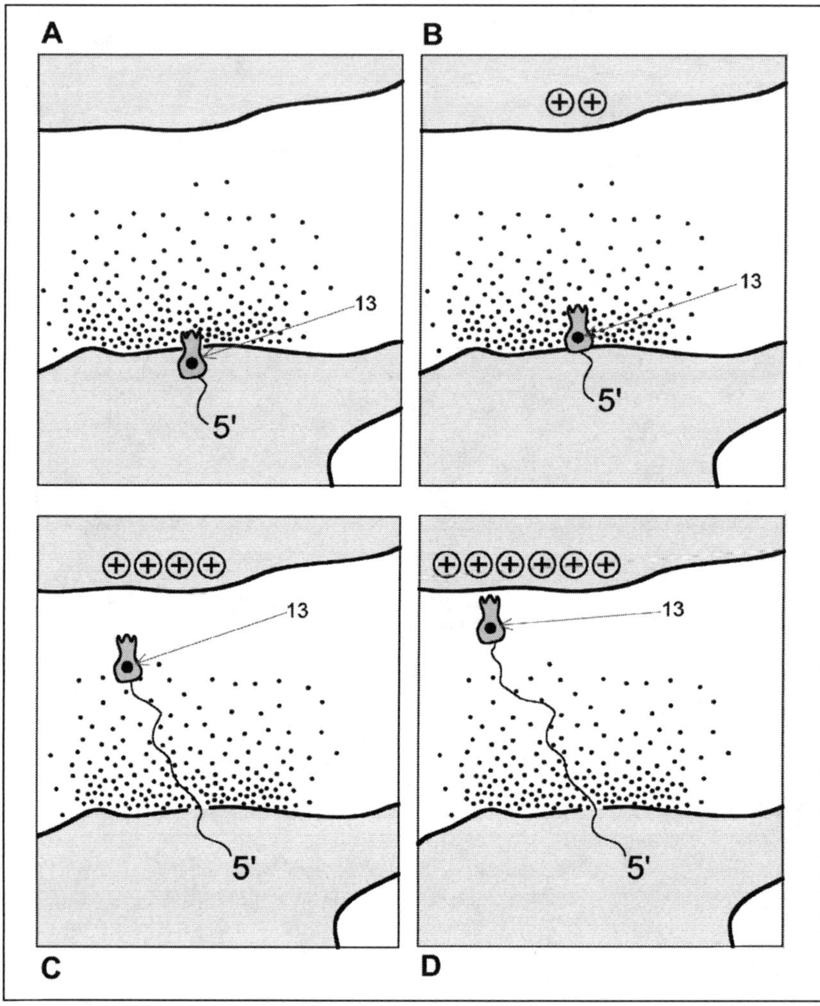

Figure 6. Locations of *Hoxa13*: A) inside CT, B) just over the surface of CT, C) at the outer border of TF area, and D) beyond the TF area.

possible explanation of this failure is that, at st21, proximal cells in the bud are not yet competent to respond to signals for *Hoxa13* expression. An experiment in progress, when completed, will test the validity of this explanation.[13]

Conclusions and Predictions

The present model does not merely describe the collinearity data but its aim is to explain the origin of collinearity in terms of fundamental and well documented processes. Specific experiments have to determine the particular features of the mechanisms involved. In this spirit, different mechanisms cooperate at several levels: (a) At the multicellular level and depending on the case, the study of modified diffusion with or without endocytosis has clarified how morphogen gradients can be established.[17-19,21] (b) Signal transduction has been determined with accuracy leading to the production and transportation of particular molecules in specific locations inside the nucleus.[34-37] (c) At the subcellular level, physical forces are produced causing translocations of the *Hox* clusters in the nucleus and as a plausible candidate an electric force (attractive or repulsive) was analyzed in more detail.[12,13] The subsequent results of gene transcription in murine ES cells not only are compatible with this hypothesis but, at the same time, they specify some features of the attractive force involved.[14] Furthermore, some recent experiments in vivo at different stages of embryonic mouse development[44] confirm the above findings and provide further support to the present model. It is still possible that, in addition to the attraction, a repulsive force may also act on the posterior end of the cluster.[12,13] However, the origin of the force remains still undetermined and a direct experiment is crucial for its specification.

To the above end, a change of the electric charge of the cluster and the consequent modification of the expression pattern would be an undisputable test of the electric hypothesis. The combination of genomic rearrangement techniques[45] with the insertion methods of charged oligonucleotides in the genome[46] could eventually serve this purpose. Accordingly, if negative charges are inserted at the 3' end of *Hoxb* cluster (increasing the total negative charge of the cluster) the pulling electrophoretic force will increase and the expressions of *Hoxb* genes will be shifted anteriorly (to the right in Fig. 1B). On the contrary, if positive charges are inserted the expressions will be shifted to the left toward the posterior end of the embryo axis.

In the present framework collinearity is the outcome of cooperation of physical principles and biochemical mechanisms. This formulation allows a unification of posterior prevalence of expressions along the embryo axis and quantitative collinearity of 5' gene expressions in the limb since, in both cases, *Hox* gene transcription obeys the same rules. In Figure 5 the distance between L and H determines a 'window between thresholds' for *Hoxa13* gene expression. This is associated to the fiber stretching inside the TF area (Fig. 4) and corresponds to a 'window of gene expression' contained between the spatial boundaries (proximal and distal). In this respect, the difference between partially overlapping and nested expression domains is only accidental: the 3' genes in the primary embryo axis cannot be accommodated in the TF area and for every cell only a subset of these gene expressions fits in. In the limb bud, when extruded in the ICD, the 5' genes can be contained in the TF area and they form a nested pattern. In both cases the posterior gene expression dominates compared to the effect of the overlapping more anterior gene expressions. Furthermore, an FGF4 bead inserted distally in the bud will increase the morphogen level above H at the tip area (Fig. 5) and accordingly *Hoxa13* gene will be shifted beyond the window of gene expressions (Fig. 6). The verification of the gradual *Hoxa13* attenuation distally[23] indicates that, as in the primary axis, the nested expression pattern can become partially overlapping as a result of appropriate gene translocations.

If the electric hypothesis proves correct, many specific questions could be asked: how reversible is the decondensed *Hox* fiber toward the initial condensed state inside the CT when the gene activation is turned off? In this spirit, a regression of *Hoxb1* was noticed from the ICD toward the CT at later stages of activation.[14]

There are some experiments where the collinearity picture is violated. For instance when the anterior *Hoxb1* is ectopically relocated at the posterior end of the *HoxD* cluster near *Hoxd13* the following expressions were observed:[47] normally *Hoxb1* is expressed in the fourth rhombomere of the developing nervous system. The expression of the relocated *Hoxb1* transgene is totally abolished in agreement with the present model since the posterior *Hoxd* genes are not expressed in this rhombomere. However the transgene expression in the mesoderm was unexpectedly not suppressed. This probably indicates that the normal collinearity behaviour depends on the developing stage, the tissue and the transient structure of the chromatin architecture. Thus, it seems that in the early mesoderm the chromatin is not dense enough and the factors of *Hoxb1* activation can recognize the ectopic transgene inducing a local opening in the neighborhood of *Hoxd13*.[47] In the same article the authors describe some specific mechanisms to this end.[47]

In *Drosophila* the developmental fate of the domains along the anterior-posterior axis is specified by the repression of the homeotic gene expressions.[48] Apart from the obvious differences from the present formulation, the ABD-B protein spreads across several parasegments in the posterior- anterior direction in a manner 'vaguely reminiscent' of the spreading of the posterior *Hox* genes (Abd-B-like genes) in the developing vertebrate limb.[6,48]

Physical forces are involved in many developmental processes.[49] In the case of *Hox* gene collinearity I think it is impossible to explain the observed regularities without using a physical principle, for the following reason: as exposed above, the chemical (molecular) cue of positional information is transformed into a chemical signal at the chromatin level. One biochemical response to this signaling is the high increase in histone H3 acetylation and methylation.[14] However, these histone modifications are not sufficient to produce the *HoxB* cluster decondensation which is the necessary first step for gene activation.[14] Only physical forces can cause such mechanical modifications of large molecular complexes (e.g., decondensation, stretching, translocation, shearing etc) and the observed collinearity reflects the correlation between the measure of the applied force and the degree of chromatin (cluster) deformation. The explanation of collinearity therefore should be based on appropriate physical mechanisms generating suitable forces acting on the nuclear macromolecules. Gene enhancers, inhibitors, cis-regulators and the other tools of the biochemical machinery, although crucial, cannot explain only by themselves all aspects of collinearity.[13] This inability and the accumulation of new complicated data led to the assumption that no single and universal mechanistic explanation may exist for these correlated phenomena.[50] Instead, tinkering without underlying logic other than the attainment of the final goal might be responsible for the observed characteristic gene expressions. In contrast, according to the present formulation, collinearity is unique and universal since it is the result of physical laws. However, in every patterning pathway along axes, the backbone of collinearity is covered by the features of the particular geometry (macroscopic and microscopic) combined with the molecular specificity of the by-side genetic activation. For example, collinearity in the primary vertebrate axis generates partially overlapping expressions while in the limb bud the expressions form a nested pattern (Fig. 1B). In other cases, the chromosome environment of the cluster does not fulfill the prerequisites for the proper appearance of *Hox* gene collinearity.[47] Whenever the universal phenomenon of collinearity is at work, the phenotypes in each case are modulated by geometry and the concurrent molecular and genetic processes. Future experiments will clarify whether tinkering or the 'dressed' core of collinearity is responsible for the observed variety of axial developmental events.

In a recent publication,[51] serial *Hoxd* gene deletions and duplications were reported. These authors propose two independent collinear mechanisms (a 'two-wave model') in order to describe the formation of proximal and distal structures in the limb bud. The general conclusion from their new experiment is that, for nested internal deletions and duplications, the expression modes (temporal and spatial) of the remaining more posterior (5') genes are adjusted in correlation with their distance from the origin- the telomeric end (3') of the cluster. These modified gene expressions are in agreement with the present model of a single and universal collinearity according to which the above distance from the origin (3') measures the length of the predicted chromatin fiber translocation in ICD. Such fiber extrusions were confirmed by Chambeyron and Bickmore.[14] The posterior end of the cluster (5') remains anchored inside the CT (Fig. 3). I think it is self-evident that physical forces cause the gradual shifts of the *Hox* genes in ICD. Therefore, it should be interesting first to verify the existence of these forces and then to explore their mode of action.

Acknowledgment

The author thanks the Editorial Board of The International Journal of Developmental Biology for granting the permission to reproduce this article in the present book.

References

1. Lewis EB. A gene complex controlling segmentation in Drosophila. Nature 1978; 276:565-570.
2. McGinnis W, Krumlauf R. Homeobox genes and axial patterning. Cell 1992; 68:283-302.
3. Krumlauf R. Hox genes in vertebrate development. Cell 1994; 78:191-201.

4. Duboule D, Dollé P. The structural and functional organization of the murine HOX gene family resembles that of Drosophila homeotic genes. EMBO J 1989; 8:1497-1505.
5. Graham A, Papalopulou N, Krumlauf R. The murine and Drosophila homeobox gene complexes have common features of organization and expression. Cell 1989; 57:367-378.
6. Dollé P, Izpisua-Belmonte JC, Falkenstein H et al. Coordinate expression of the murine Hox-5 complex homeobox-containing genes during limb pattern formation. Nature 1989; 342:767-772.
7. Yokouchi Y, Sasaki H, Kuroiwa A. Homeobox gene expression correlated with the bifurcation process of limb cartilage development. Nature 1991; 353:443-445.
8. Nelson CE, Morgan BA, Burke AC et al. Analysis of Hox gene expression in the chick limb bud. Development 1996; 122:1449-1466.
9. Izpisua-Belmonte JC, Falkenstein H, Dollé P et al. Murine genes related to the Drosophila AbdB homeotic gene are sequentially expressed during development of the posterior part of the body. EMBO J 1991; 10:2279-2289.
10. Dollé P, Izpisua-Belmonte JC, Brown JM et al. HOX-4 genes and the morphogenesis of mammalian genitalia. Genes and Dev 1991; 5:1767-1777.
11. Kmita M, Fraudeau N, Hérault Y et al. Serial deletions and duplications suggest a mechanism for the collinearity of Hoxd genes in limbs. Nature 2002; 420:145-150.
12. Papageorgiou S. A physical force may expose Hox genes to express in a morphogenetic density gradient. Bull Math Biol 2001; 63:185-200.
13. Papageorgiou S. A cluster translocation model may explain the collinearity of Hox gene expressions. BioEssays 2004; 26:189-195.
14. Chambeyron S, Bickmore WA. Chromatin decondensation and nuclear reorganization of the HoxB locus upon induction of transcription. Genes and Dev 2004; 18:1119-1130.
15. Wolpert L. One hundred years of positional information. Trends Genet 1996; 12:359-364.
16. Summerbell D, Lewis JH, Wolpert L. Positional information in chick limb morphogenesis. Nature 1973; 244:492-496.
17. Green J. Morphogen gradients, positional information and Xenopous: Interplay of theory and experiment. Dev Dyn 2002; 225:392-408.
18. Scholpp S, Brand M. Endocytosis controls spreading and effective signaling range of Fgf8 protein. Curr Biol 2004; 14:1834-1841.
19. Belenkaya TY, Han C, Yan D et al. Drosophila Dpp morphogen movement is independent of dynamin-mediated endocytosis but regulated by the glypican members of heparan sulfate proteoglycans. Cell 2004; 119:231-244.
20. Kerszberg M, Wolpert L. Mechanisms for signaling by morphogen transport: A theoretical study. J Theor Biol 1998; 191:103-114.
21. Smith JC, Gurdon JB. Many ways to make a gradient. BioEssays 2004; 26:705-706.
22. Papageorgiou S. Cooperating morphogens control hoxd gene expression in the developing vertebrate limb. J Theor Biol 1998; 192:43-53.
23. Vargesson N, Kostakopoulou K, Drossopoulou G et al. Characterization of Hoxa gene expression in the chick limb bud in response to FGF. Dev Dyn 2001; 220:87-90.
24. Papageorgiou S, Almirantis Y. Gradient model describes the spatial-temporal expression pattern of Hoxa genes in the developing vertebrate limb. Dev Dyn 1996; 207:461-469.
25. Belmont AS, Bruce K. Visualization of G1 chromosomes: A folded, twisted, supercoiled chromonema model of interphase chromatid structure. J Cell Biol 1994; 127:287-302.
26. Tumbar T, Sudlow G, Belmont AS. Large-scale chromatin unfolding and remodeling induced by VP16 acid activation domain. J Cell Biol 1999; 145:1341-1354.
27. Cremer T, Cremer C. Chromosome territories, nuclear architecture and gene regulation in mammalian cells. Nat Rev Genet 2001; 2:292-301.
28. Schlossherr J, Eggert H, Paro R et al. Gene inactivation in Drosophila mediated by polycomb gene product or by position-effect variegation does not involve major changes in the accessibility of the chromatin fibre. Mol Gen Genet 1994; 243:453-462.
29. Volpi EV, Chevret E, Jones T et al. Large-scale chromatin organization of the major histocompatibility complex and other regions of human chromosome 6 and its response to interferon in interphase nuclei. J Cell Sci 2000; 113:1565-1576.
30. Cheutin Y, McNairn AJ, Jenuwein T et al. Maintenance of stable heterochromatin domains by dynamic HP1 binding. Science 2003; 299:721-725.
31. Van Driel R, Fransz PF, Verschure PJ. The eukaryotic genome: A system regulated at different hierarchical levels. J Cell Sci 2003; 116:4067-4075.
32. Visser AE, Jaunin F, Fakan S et al. High resolution analysis of interphase chromosome domains. J Cell Sci 2000; 113:2585-2593.

33. Marg A, Shan Y, Meyer T et al. Nucleocytoplasmic shuttling by nucleoporins Nup153 and Nup214 and CRM1-dependent nuclear export control the subcellular distribution of latent Stat1. J Cell Biol 2004; 165:823-833.
34. Bourillot PY, Garrett N, Gurdon JB. A changing morphogen gradient is interpreted by continuous transduction flow. Development 2002; 129:2167-2180.
35. Shimizu K, Gurdon JB. A quantitative analysis of signal transduction from activin receptor to nucleus and its relevance to morphogen gradient interpretation. Proc Natl Acad Sci USA 1999; 96:6791-6796.
36. Pan D, Estevez-Salmeron LD, Stroschein SL et al. The integral inner nuclear membrane protein MAN1 physically interacts with the R-Smad proteins to repress signalling by the transforming growth factor-{beta} superfamily of cytokines. J Biol Chem 2005; 280:15992-16001.
37. Itoh K, Brott BK, Ratcliffe MJ et al. Nuclear localization is required for Dishevelled function in Wnt/beta-catenin signalling. J Biol 2005; 4(1):3.
38. Cui Y, Bustamante C. Pulling a single chromatin fiber reveals the forces that maintain its higher-order structure. Proc Natl Acad Sci USA 2000; 97:127-132.
39. Stigter D, Bustamante C. Theory for the hydrodynamic and electrophoretic stretch of tethered B-DNA. Biophys J 1998; 57:1197-1210.
40. Ferree S, Blanch HW. Electrokinetic stretching of tethered DNA. Biophys J 2003; 85:2539-2546.
41. Long D, Viovy J, Ajdari A. Stretching DNA with electric fields revisited. Biopolymers 1996; 39:755-759.
42. Ferree S, Blanch HW. The hydrodynamics of DNA electrophoretic stretching and relaxation in a polymer solution. Biophys J 2004; 87:468-475.
43. Katritch V, Bustamante C, Olson WK. Pulling chromatin fibers: Computer simulations of direct physical micromanipulations. J Mol Biol 2000; 295:29 40.
44. Chambeyron S, Da Silva NR, Lawson KA et al. Nuclear reorganisation of the Hoxb complex during mouse embryonic development. Development 2005; 132:2215-2223.
45. Herault Y, Rassoulzadegan M, Cuzin F et al. Engineering chromosomes in mice through targeted meiotic recombination (TAMERE). Nat Genet 1998; 20:381-384.
46. Pitard B , Bello-Rroufai M, Lambert O et al. Negatively charged self-assembling DNA/poloxamine nanospheres for in vivo gene transfer. Nucleic Acids Res 2004; 32(20):e159.
47. Kmita M, van der Hoeven F, Zakany J et al. Mechanisms of Hox gene colinearity: Transposition of the anterior Hoxb1 gene into the posterior HoxD complex. Genes and Dev 2000; 14:198-211.
48. Busturia A, Bienz M. Silencers in Abdominal-B, a homeotic Drosophila gene. EMBO J 1993; 12:1415-1425.
49. Gordon R. The hierarchical genome and differentiation waves. London: World Scientific, 1999.
50. Kmita M, Duboule D. Organizing axes in time and space; 25 years of colinear tinkering. Science 2003; 301:331-333.
51. Tarchini B, Duboule D. Control of Hoxd Genes' collinearity during early limb development. Developmental Cell 2006; 10:93-103.

Index

H

I

K

L

M

N

O

P